Reshetnev Readings 2019

IOP Conference Series: Materials Science and Engineering
Volume 1230

Krasnoyarsk, Russia
11 – 15 November 2019

Editors:

Edkham Sh. Akbulatov
Michael Affenzeller
Stefan Wagner
Aleksandr F. Gavrilov
Nikolai A. Testoedov
Rinat G. Galeev
Yuri Yu. Loginov
Ivo Kraus
Xu Huibin
Vasilii F. Shabanov

ISBN: 978-1-7138-4688-8
ISSN: 1757-8981

Printed from e-media with permission by:

Curran Associates, Inc.
57 Morehouse Lane
Red Hook, NY 12571

Some format issues inherent in the e-media version may also appear in this print version.

Institute of Physics
Dirac House, Temple Back
Bristol BS1 6BE UK

Phone: 44 1 17 929 7481
Fax: 44 1 17 920 0979

techtracking@iop.org

Additional copies of this publication are available from:

Curran Associates, Inc.
57 Morehouse Lane
Red Hook, NY 12571 USA
Phone: 845-758-0400
Fax: 845-758-2633
Email: curran@proceedings.com
Web: www.proceedings.com

TABLE OF CONTENTS

23nd International Scientific Conference

Reshetnev Readings-2019 (11–15 November 2019)

Preface to 23nd edition of Reshetnev Readings Conference Proceedings

The Russian Academician Mikhail Reshetnev (1924–1996) is one of the pioneers of the Soviet space program. His contributions were especially important to Soviet (later, Russian) satellite communication and navigation systems. As a sign of gratitude for Reshetnev's achievements, two institutions – our University and a satellite manufacturing company were named in his honor. 2019 is the 23rd time that the University has organized a scientific conference in memory of Academician Reshetnev.

Prior to 2017, the University was named Reshetnev Siberian State Aerospace University. After it merged with the Siberian University of Technology in 2016, its name was changed to Reshetnev Siberian State University of Science and Technology, or Reshetnev University. 2017 is, therefore, the first time that the University is holding the Conference in a new status. While the traditions sections of the Conference with the fields of aerospace technology and engineering, computer science and telecommunication systems remain as the nucleus of the Conference, new sections, including a significantly developed section or workshop on Materials Science and Engineering have been added. The scope of the section has been broadened, adding new areas of research such as forestry products and biomaterials in engineering, solid-state physics, composite materials, polymers, and other areas related to the science of developing and producing new materials. The Conference is traditionally held in the Siberian city of Krasnoyarsk. Over 200 participants annually take part in the Conference, making it one of the largest scientific events in the city.

In 2019, the topic of the Conference was expanded to house a growing demand for research reports in the field of Earth and environmental studies. The majority of these topics are related to remote sensing, geographical information systems, environmental issues, and ecology.

The Conference is multidisciplinary. However, the topics of the reports should be relevant to engineering and industry.

Pr. Yurii Loginov

Vice Rector for Science and Research Reshetnev University

Editors

Edkham Sh. Akbulatov, Rector, Reshetnev Siberian State University of Science and Technology, Krasnoyarsk, Russian Federation

Michael Affenzeller, Vice Dean for R&D School of Informatics, Communications and Media University of Applied Sciences Upper Austria, Linz, Austria

Stefan Wagner, Professor for Complex Software Systems · University of Applied Sciences Upper Austria, Linz, Austria

Aleksandr F. Gavrilov, CEO, JSC Krasnoyarsk Machine-Building Plant, Krasnoyarsk, Russian Federation

Rinat G. Galeev, CEO JSC NPP Radiosviaz, Krasnoyarsk, Russian Federation

Yuri Yu. Loginov, Vive Rector for Research, Reshetnev Siberian State University of Science and Technology, Krasnoyarsk, Russian Federation

Ivo Kraus, Professor, Czech Technical University in Prague, Prague, Czech Republic

Xu Huibin, President, Beihang University, Beijing, China

Nikolai A. Testoedov, CEO, JSC Academician M.F. Reshetnev Information Satellite Systems, Zheleznogorsk, Russian Federation

Vasilii F. Shabanov, Scientific Supervisor, Federal Research Center Krasnoyarsk Science Center of the Siberian Branch of the Russian Academy of Sciences, Krasnoyarsk, Russian Federation

Topics

1. Aerospace Engineering. Issues of Design and Production.
2. Thermodynamics of Industrial Processes.
3. Design and Exploitation of Spacecraft.
4. Design and Exploitation of Rocket Engines.
5. Control Systems and Space Electronics.
6. Robotics and Mechatronics Engineering.
7. Strength of Materials.
8. Production of Materials (Natural and Synthetic).
9. Machining.
10. Single Crystals.
11. Welding Technologies.

12. Materials Science.

12. Composite Materials.

13. Chemistry.

IOP Conf. Series: Materials Science and Engineering 1230 (2022) 011002 doi:10.1088/1757-899X/1230/1/011002

Peer review declaration

All papers published in this volume of IOP Conference Series: Materials Science and Engineering have been peer reviewed through processes administered by the Editors. Reviews were conducted by expert referees to the professional and scientific standards expected of a proceedings journal published by IOP Publishing.

- **Type of peer review:**
  Single-blind. All papers are sent to a committee of anonymous reviews, where they are selected according to subject area. The paper's quality is assessed by these reviewers.

- **Conference submission management system:**
  Conference papers are submitted though the conference website.

- **Number of submissions received: 126**

- **Number of submissions sent for review: 87**

- **Number of submissions accepted: 54**

- **Acceptance Rate (Number of Submissions Accepted / Number of Submissions Received X 100): 62%**

- **Average number of reviews per paper: 1**

- **Total number of reviewers involved: 12**

- **Any additional info on review process: none**

- **Contact person for queries:**
  Name : Prof. Yuri Yu. Loginov
  Affiliation: Reshetnev Siberian State University of Science and Technology
  Email : loginov@sibsau.ru

Influence of the boundary effect, viscosity factor and heat conductivity mechanism on the shock front evolution

A L Adrianov

Reshetnev Siberian State University of Science and Technology, 31 Krasnoyarskii rabochii prospekt, Krasnoyarsk, 660037, Russia
Siberian Federal University, 79 Svobodny Av., 660041 Krasnoyarsk, Russia

E-mail: adrian58al@gmail.com

Abstract. In this paper a new economical discrete-analytical approach is applied to the problem penetrating of a shock wave in the shear layer of a viscous heat-conducting gas at high Reynolds numbers.

1. Introduction

Mathematical modeling of *shock flows* of an inviscid as well as viscous heat-conducting gas at high Reynolds numbers (Re_∞), based on the approaches combining the advantages of numerical and analytical methods is obviously promising. If constructed correctly, a *numeric and analytical method-symbiosis* can have much greater resolving power and/or efficiency in calculating gasdynamic flows characterized by *shock waves*, *shear layers*, as well as their interactions than some of those belonging to the classes of numerical and analytical methods. The methods based on the above mentioned symbiosis are able to give new information concerning the details of the inner structure of complex two dimensional, in particular, stationary *gasdynamic shock flows*; these methods can also reveal both individual and combined impact of such factors as the *non-uniformity* of an undisturbed flow in front of the shock wave, the *boundary effect* behind it which is formed by disturbances catching up with it, and *viscosity-heat conductivity* on the physical process under study. Here, the term "*boundary effect*" is understood as being a derivative from the generally recognized term "*boundary condition*" when applied to the back surface of the shock wave; the boundary effect is associated with a continuous system of the corresponding conservation laws. (The analytical apparatus in such *numerical analytical methods* is applied due to the *schematization* (isolation) of the shock front by the flat curvilinear surface of a *strong gasdynamic discontinuity*; thus, along with the known relations in the case of the inclined shock wave (or, even, instead of them), use is made of the continuous-*generalized differential relations* asymptotically taking into account the *viscosity factor* and *mechanism of heat conductivity* (VHC factor) [1–5]. Here, it is worth mentioning the contribution of V.N. Uskov who paid great attention to the analytical apparatus in the research of shock wave flows. This creative process was continued by his students, in particular, by the author of the present study who has attempted to combine the advantages of the previously distinct analytical and numerical methods.

2. Discrete-analytical approach

In the case of numerical modeling of *shock flows* of a viscous heat-conducting gas with large Re_∞ numbers without the shock wave *schematization*, performing the "through" calculations it is almost impossible to distinguish the contribution into the gasdynamic parameters realized behind it due to the action of the external (!) (from the side of the macro-flow on both sides of the shock wave outside the shock transition), viscous stresses and heat flow [1–3], in other words, to correctly switch off the action of the external VHC factor only on the shock wave, thereby revealing the contribution of this factor in the shock process. Here and in what follows it is not schematic (numeric) but physical external *viscosity factor* and *conductivity mechanism* that are implied, which do not have to be taken into account within the method of the through calculations of *shock flows*. The main advantages of the *shock capturing method* (SCM) include its versatility, relative simplicity of constructing a calculation algorithm (its software implementation) and, as a consequence, some reliability. The actual accuracy (without taking into account the unavoidable approximate losses on the shocks), obtained using SCM in most cases is satisfactory for practical applications; thus, this direction is intensively being developed.

In all fairness, it is worth noting that good fulfillment of the *standard conditions on the inclined (straight)* shock wave often means that at high Re_∞ the VHC factor on the *strong gasdynamic discontinuity* can be neglected; in the integral formulas of the schematized shock transition this factor is used with the small parameter; the gradients on both sides of the shock wave are assumed to be limited [1]. In the case of a uniform flow on both sides of the shock wave the contribution of the above mentioned factor into the parameters of the shock transition is generally absent at any Re_∞, since the external gradients of the values are absent and the shock itself is smooth.

If the shock wave *is schematized*, then, it is obvious that in the general case of the non-uniform macroflow outside its curvilinear surface, the influence of the VHC factor on the parameters of the shock transition must correctly be taken into account by the mathematical model even at the average values of ($\sim 10^3$) Re_∞, which results into the necessity to use generalized *conditions* on such a shock wave. However, in the case of *schematization* of the shock wave, in terms of the boundary problem statement (!) it appears to be impossible to vary all the possible boundary effects in a wide range, to reveal the degree of their influence on the shape of the shock itself, as well as on the distribution of the parameters and their derivatives along its back surface and the action of the VHC factor. Besides, the *numerical* methods for solving the systems of *Euler equations and Navier-Stokes equations of viscous heat conducting gas* (NSEVHC) integrated into a particular commercial software, still require considerable calculation resources, which cannot be neglected despite the existence of supercomputers with petaflop performance.

The above mentioned problems which appear in the numerical modeling of *shock flows* turn out to be surmountable by using the *numerical-analytical methods*, and, in particular, the "*discrete-analytical approach*" [1].

Let us specify what is implied by the compound term *numerical-analytical method* (as well as by the terms *difference-analytical method, discrete-analytical approach).* In the general case, this is not to be understood as a method constructed based on direct (independent) combining of the numeric (discrete) and analytical apparatus which allows constructing some "total solution" of a particular problem. However, in some cases *an analytical* (in the extended sense [6]) solution for the *shock process* is constructed by its "submersion" into another solution-source which can be determined by analytical dependences or have a numerical origin – the gasdynamic field obtained from the preliminary calculation and tabularized using smooth interpolation. A possibility of obtaining analytical solutions in the classical sense is often connected with the assumptions simplifying the problem statement, consideration of particular cases etc. There are numerous studies dealing with such methods since in the pre-computer era and even with the appearance of the first low-performance PCs these methods and the physical experiment were the main *tools* for studying complex aero gas dynamic processes. These methods have been used and improved until now since they prove to be efficient in the mathematical description of particular high-gradient *shockwave* and *shock-free* gasdynamic processes. In contrast, in the studies being carried out only some minimum assumptions are made which allow, as a maximum, the reduction of the boundary problem for the calculated *shock flow* in terms of the Euler equations or NSEVHC to *the Cauchy problem* for a nonlinear ODE system which then, is solved numerically. The final *working equations* of this system are obtained by complex mathematical calculations by means of computer algebra [7]: they are exact only in terms of gas dynamics; however, taking into account the viscosity - heat conductivity factor, they are approximate asymptotic equations containing the elements of difference approximations. Unfortunately, such an ODE system requires the development of non-standard approaches for its numerical integration. [1, 3]. In any case, *analytical or exact solutions* can be discussed only in the *extended sense* [6].

To conclude, *the numerical-analytical method* in the present study is understood to be *a method symbiosis* constructed using the synthesis of the apparatus of difference schemes and analytical apparatus applied locally (*discretely*) to particular characteristics. In this case, the shock wave appears to be either *distinguished*, or *schematized* by the flat curvilinear surface of the *strong gasdynamic discontinuity,* depending on the applied mathematical model of the compressible gas flow: inviscid (ideal) or viscous with heat conductivity at high Re_∞, correspondingly. The main motivation for the *smooth representation* of the shock wave with the adequate local application of the analytical apparatus is the natural (infinite (!) in the absence of other disturbances, and, moreover, taking into account the VHC factor) order of the smoothness of the solution in the direction tangent to the shock, which is to be used. Here, if on the *schematized* curved shock wave in a non-uniform flow of viscous or inviscid heat-conducting gas, the *relations of the 0-th order* hold, it is required that the corresponding "continuous relations", *relations of the 1-st order* (in other words, *differential relations* in the shock wave [1–2]) should be fulfilled. Using the mathematical language of spline apparatus (spline functions) one can express this idea in a different manner: when used locally, the analytical apparatus allows passing to the spline of a higher order in *the smooth representation* of the shock wave front. It is important that in the case of such a *smooth representation* of the shock wave there is no necessity of point-to-point statement of the *0-th order relations*: in any of the calculation points they will be fulfilled automatically (!), and, thus, can be employed in the course of the calculation only for the purpose of additional point-to-point control of calculation errors. The analytical apparatus of *differential relations* on the shock wave employed in connection with the smoothness of the process allows one to analyze the influence (in particular, the combined one) of the main physical parameters on the evolution of the shock front itself and to control its behavior. These factors are the following: gradients of gasdynamic values in an undisturbed flow in front of the shock wave, boundary effect behind it, as well as VHC factor when it is asymptotically taken into account in the present relations. Note that based only on the *standard* (of the Rankine-Hugoniot type) *relations of the 0-th order* and, correspondingly, on the apparatus of the *shock* or *shock wave polars* [1] (traditionally used by experimentalists and other specialists in calculations at present) such an analysis and control of the behavior of the shock wave front cannot be made.

The motivation for constructing effective *numerical-analytical methods* for the calculation of *shock flows* of inviscid and, in particular, of viscous heat-conducting gas at high Re_∞ has a more interesting mathematical background. The shock wave when it has a pronounced (discontinuous upon *schematization*) change of gasdynamic values in its small neighborhood, and has "the margin of smoothness" in the tangential direction which, undoubtedly (!), would be relevant for the calculation algorithm of the problem solving. Otherwise, as is often the case when, for example, difference SCM are used, apart from the usual local loss of accuracy for the shock wave and other approximative defects, the so-called *defect of saturation* of the calculation method will appear. [8]. The essence of this phenomenon which is negative but less pronounced than the approximation error, is in the fact that beginning from some critical point the next gradual rank increase of the solution smoothness no longer results in the adequate decrease of the calculation error of the algorithm, i.e. it is no longer meaningful. In particular, this brings about the necessity of the local application of the analytical apparatus in the calculations of gasdynamic flows containing shock waves or other special features. Moreover, if a shock wave is significantly isolated from the oncoming and catching up neighboring features or the problem is solved in a more general case with respect to the small VHC factor automatically providing a greater smoothness of the solution (!), the role of the above mentioned shock increases. For example, in *schematizing* a single shock wave penetrating in *the shear layer* (*layer*) of a viscous heat-conducting gas at high Re_∞, the initial initial-boundary problem for NSEVHC and moreover, in terms of its inviscid approximation for Euler equations when the *layer* is considered to be a *vortex layer* with a non-degenerate profile of entropy, velocity and enthalpy, the reduction to the Cauchy problem for the ODE system becomes possible.

The above mentioned reduction of NSEVHC to ODE in terms of the statement of the viscous problem in the process of the mathematical modeling within a unified computational algorithm, on the one hand, allows one to pass "through" from the *gasdynamic* to the *diffusion* stage of the shock wave evolution in the *layer*, given different boundary effects, while on the other hand, it enables one to significantly save computational power: the calculation on PC takes several seconds rather than numerous hours (NSEVHC). Here, as indicated, in such problem statement the natural smoothness of the solution (especially when the VHC factor is taken into account) in the direction tangent to the shock wave is retained, which was the main motivation for the development and application of the *discrete-analytical approach* [1].

3. Generalized differential relations

Consider the given problem of the shock wave penetration into the *layer* of the inviscid or viscous heat-conducting gas at high Reynolds numbers. On the surface of the *schematized* curved shock wave either ordinary or *generalized relations* [1–3] of the 0-th and 1-st (differential) order asymptotically taking into account the effect of the viscosity - heat conductivity factor hold; they are exact in terms of gas dynamics. In deriving *generalized differential relations* (GDR) a number of assumptions is used [1–3], including the assumption on the scale of derivatives of the main values in the *layer*. The asymptotic order of the VHC terms preserved in GDR corresponds to $O(1)$ at $\mathrm{Re}_\infty \to \infty$, which appears to be quite sufficient for the VHC effect to be estimated. The compact matrix record of GDR [1, 2] fulfilled in such a shock wave can be written as:

$$a_{ij}\left(\hat{\Phi}_j\right)_{\hat{n}} + b_i K_w + c_{ij}\left(\Phi_j\right)_n + d_i\frac{\delta}{y} + \varepsilon^2\left[e_{ij}\left(\hat{\Phi}_j\right)_{\hat{n}\hat{n}} + f_i K_w{}' + g_{ij}\left(\Phi_j\right)_{nn} + N_i\right] = 0,$$
$$\Phi = \left(W, \theta, p, h\right)^T; \qquad i, j = 1, \ldots, 4, \tag{1}$$

where $\delta = 0$ $(\delta = 1)$; $\varepsilon^2 = \left(\mathrm{Re}_\infty\right)^{-1}$; $N_i = N_i\left(\left(\hat{\Phi}_j\right)_{\hat{n}}, \left(\Phi_j\right)_n, \hat{\Phi}_j, \Phi_j, K_w\right)$.

In (1) the summation over the repeated index j is applied; the values K_w and $K_w{}' = \dfrac{d^2\Omega}{dw^2}$, where $\Omega = \theta + \sigma \equiv \hat{\theta} + \hat{\sigma}$, represent the longitudinal curvature of the shock wave (as opposed to radial curvature $1/y$ in the axisymmetric case) and its derivative; the vector **N** explicitly combines nonlinear terms (products of lower derivatives and curvature). All the values are assumed to be dimensionless. In (1) in the notations K_w and $K_w{}'$ $w \equiv \tau_s$ – is the coordinate in the direction which is longitudinal to the shock wave (in contrast to the velocity module W it is denoted by the small w); p is the pressure; ρ is the density; h is the enthalpy; θ and $\hat{\theta}$ are the angles of inclination of the streamline to the axis OX of the Cartesian (cylindrical) system of coordinates XOY in front of and behind the shock wave, respectively; all the derivatives are presented in the local system of *natural (eigen-) coordinates* (s, n) in front of the shock wave, $(\hat{s}, \hat{n})$ – behind the shock wave; the derivatives in front of the shock wave are assumed to be known. The acute angle σ between the velocity vector $\vec{w}$ and the shock wave (between $\vec{s}$ and $\vec{\tau}_s$) determines the shock wave intensity ($J = \hat{p}/p = (1+\varepsilon)\mathrm{M}^2\sin^2\sigma - \varepsilon$; $\varepsilon = (\gamma - 1)/(\gamma + 1)$; $\gamma = c_p/c_v$): $|\sigma| \geq \alpha_\mathrm{M}$, where $\alpha_\mathrm{M} = \arcsin\left(1/\mathrm{M}\right)$ is the *Mach angle* at a given point in front of the shock wave and the sign χ of this angle determines the family to which the shock wave belongs (the characteristic, at $\sigma = \chi\alpha_\mathrm{M}$) in the inviscid case. The functional matrices (vectors) of the coefficients **A**,…, **G** in (1) in a complex way depend on the gasdynamic values on both sides of the shock wave; in the inviscid case ($\varepsilon^2 = 0$ в (1)) **A**,…, **D** after the corresponding admissible transformations exactly results in analogous coefficients in [4, 5], where Uskov V.N. used another, more convenient in this case, group of dependent variables.

Complex analytical *machine* calculations were performed in the symbol transformation system «REDUCE» on PC in order to obtain the symbolic products to be used later in *calculation* software modules, namely *generalized relations of the 0–th* and *1-st* (GDR (1)) *order* on the curved shock wave, and those following from (1), after their closure and difference approximation of the upper derivatives behind the shock wave, differential-difference *working equations*, their Jacobians etc. [7].

The application of GDR (1) implies their closure; for this it is required to set any derivative behind the shock wave or their combination, thereby determining the boundary effect behind it. As such a combination use can be made of an additional *differential constraint*, in particular, that excluding the boundary effect [1]. In the case of setting the boundary effect behind the shock wave, i.e. closing the system (1) which is nonlinear (at $\varepsilon^2 \neq 0$) in the first derivatives behind the shock wave, it becomes uniquely solvable with respect to these derivatives and shock wave curvature in any of its calculation points and can be numerically integrated along the shock wave front.

Having generalized all possible boundary effects behind the shock wave, one has the *extended differential constraint*, which allows excluding $\hat{p}_{\hat{n}}$ from (1), in the following form

$$\hat{p}_{\hat{s}} + b \cdot \hat{m}^{-\chi} \cdot \hat{p}_{\hat{n}} = c, \quad \hat{m}^{-\chi} = \frac{-\chi}{\sqrt{\hat{M}^2 - 1}} = -\chi \operatorname{tg}(\hat{\alpha}_M), \tag{2}$$

where $\hat{\alpha}_M$ is the Mach angle behind the shock wave, and b and c are the *constraint* parameters. Substituting (2) in (1) it is assumed that the tangent derivative of pressure $\hat{p}_{\hat{s}}$, is substituted through the Euler equations of the normal derivative of the inclination angle $\hat{\theta}_{\hat{n}}$. Varying the constraint coefficients (2) allows one to set *strong* and *weak* boundary effects behind the shock wave of the front compression type (intensification), rarefaction (attenuation), as well as elimination of the boundary effect (neutral boundary effect). Among the boundary effects are the forbidden ones which are defined by the Uskov angle [1], which in a wide range of the shock wave intensities for the fixed M turned out to be close to the known Mach angle (being actually «its double») behind the shock wave, but as opposed to the latter it is related only to the *problem of the first order* (differential one).

4. Results, discussion and conclusion

Based on (1)–(2), the shock wave penetrating into the *shear layer* is calculated with the initial intensity J_∞ ($t_\% = 63{,}7\,\%$), which is taken in $t_\%$ percents of the logarithm of the sound intensity J_s ($\hat{M}\big|_{J=J_s} = 1$), $J_\infty(t_\%) = J_s^{t_\%/100\%}$. As a model of undisturbed *layer* the supersonic part of the boundary layer on a heat insulated plate ($\delta = 0$) is taken; Pr = 1; the power coefficient in the viscosity law is $\omega = 1$; $\partial p/\partial n \sim 0$; $W|_{y=0} = \partial p/\partial n|_{y=0} = \partial h/\partial n|_{y=0} = 0$.

Figure 1 shows the *characteristic corridor* for *viscous* and *inviscid* solutions with different boundary effects and without them. *The characteristic corridor* presents in each calculation points of the shock wave the values $|\chi\alpha_M|$ (in front of the shock wave), $|\sigma|$ (on the shock wave), $|\beta + \chi\hat{\alpha}_M|$ (behind the shock wave), which are the modules of the shock wave inclination angles and characteristics of the same family on both sides of the streamline in the undisturbed flow.

Obviously, with the complete degeneration of the shock wave due to the boundary effect or/and viscosity - heat conductivity all the three curves of the same type from different families (see figure 1), corresponding to a particular solution would intersect at one point below: *the corridor* «collapses». As is seen in figure 1, regardless of whether the boundary effect weakens or enhances the shock wave, there will be a partial narrowing of the *characteristic corridor* owing to the weakening effect of the VHC factor on the shock wave. Thus, if there is a boundary effect weakening the shock wave, as a result of the total shockwave-weakening effect of both factors, the narrowing of the *corridor* turns out to be more significant.

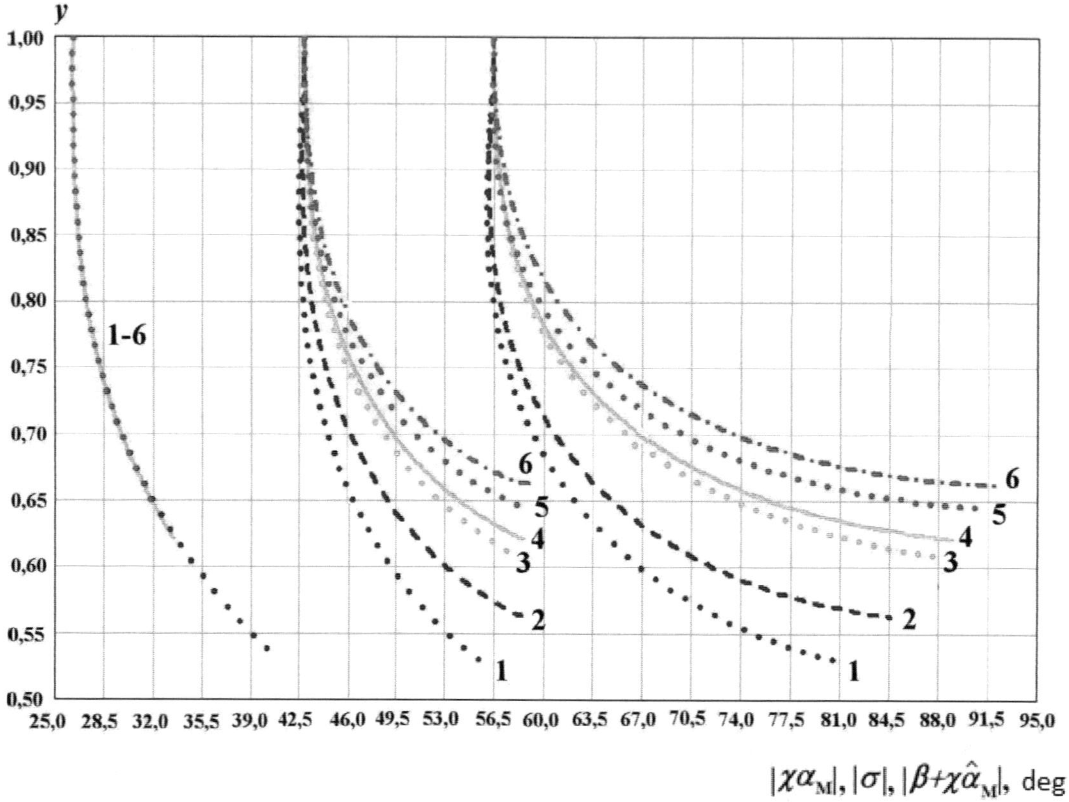

Figure 1. *Characteristic corridor*: on the left — in front of the shock wave; in the center — on the shock wave; on the right — behind the shock wave; **1** – viscous solution with the boundary effect, weakening the shockwave; **2** – inviscid solution with the boundary effect weakening the shockwave; **3** – viscous solution without the boundary effect; **4** – inviscid solution without the boundary effect; **5** – viscous solution with the boundary effect, enhancing the shockwave; **6** – inviscid solution with the boundary effect, enhancing the shockwave.

Note that the "inviscid solution" with different boundary effects presented in figure 1 means that the VHC factor was switched off ($\varepsilon^2=0$ in (1)) only in the problem of the shockwave interaction with the *layer*, i.e. the latter was considered in this case only as a *vortex layer*, though it had been formed with account of the *viscosity* and *heat conductivity* factor.

Additionally, during the computational experiment, the degree of the influence of individual parameters of the *discrete analytical model* [1], such as the order of asymptotic expansions, approximation order of difference representations, etc. on the solution has also been revealed.

Thus, an important conclusion can be made. From the studies on the shock wave penetration into an ordinary *shear layer* it follows that even with significant boundary effects of one sign or another and the fixed average (~ 10^3) number of Re_∞, taking into account the *external* viscosity - heat conductivity factor does not (!) significantly reduce the intensity of the shock wave as compared with the inviscid formulation of the problem. This is confirmed by numerous physical experiments in which these factors are always present.

In the general case, all the incoming disturbances, including strong and weak (in an inviscid formulation) discontinuities, usually occur "spontaneously", determining the further evolution of all the outcoming (initial) elements of the resulting shock wave structure, which can be optimized by solving a specific problem [5] (Uskov V.N.) [9–12].

This study considers only smooth secondary disturbances (such as compression-rarefaction) arriving at a single schematized front of the shock wave; due to the use of the *viscosity factor* and *heat conductivity mechanism*, weak discontinuities are also excluded. It is worth noting that the formation of the shock front, as well as its subsequent dynamics are always determined from two sides! Therefore, if, in addition to one or another boundary effect, the secondary smooth disturbances in front of the shock wave are also added to the problem, then the goal of possible optimization may be to maximize the process of degradation of the shock front itself. It is extremely difficult to select the necessary set of secondary disturbances, matching (synchronizing) them on both sides of the front, even in terms of a single gas model, boundary formulation of the problem, having powerful computing facilities (experimental equipment). However, in the framework of the *discrete-analytical approach* [1] such studies are actually being carried out.

Acknowledgments

To the memory of my Teacher,
Doctor of Technical Sciences,
Professor Vladimir Nikolayevich Uskov

References

[1] Adrianov A L 2016 Mathematical modeling of shock flows of ideal and viscous heat conducting gas based on the discrete-analytical approach: monograph. *Krasnoyarsk: Siberian Federal University* 216 p

[2] Adrianov A L 2009 Generalized differential relations on a shock wave. *Problems of Nuclear Science and Technology.* (Issue Mathematical modeling of physical processes) **4** pp 22–30

[3] Adrianov A L 2010 Mathematical modeling of shock flows of viscous heat conducting gas based on the asymptotic model. *Problems of Nuclear Science and Technology* (Issue Mathematical modeling of physical processes) **4** pp 10–26

[4] Uskov V N 1983 Interaction of stationary gasdynamic discontinuities. *Supersonic gas jets: Collected papers* (Novosibirsk: Nauka. Siberian Department) pp 22–46

[5] Adrianov A L, Starykh A L and Uskov V N 1995 Interaction of stationary gasdynamic discontinuities. *Novosibirsk: Nauka. Siberian Publishing house* 180 p

[6] Sidorov A F, Shapeev V P and Yanenko N N 1984 Method of differential constraints and its applications in gas dynamics (Novosibirsk: Nauka) 272 p
[7] Valiullin A N, Ganzha V G, Ilyin V P, Shapeev V P and Yanenko N N 1984 The problem of automatic construction and research of analytical difference schemes on computers *Doklady AS USSR* **275 (3)** pp 528–532
[8] Babenko K I 1986 Fundamentals of numerical analysis (M.: Science) Ch. ed. Phys.-Math. lit. 744 p
[9] Omelchenko A V and Uskov V N 1995 Optimal shock-wave systems *Bulletin of the Russian Academy of Sciences. Mechanics of fluid and gas*. **6** pp 118-126
[10] Uskov V N and Chernyshov M V 2006 Special and extreme triple configurations of shock waves *Applied Mechanics and Technical Physics* **47(4)** pp 39-53
[11] Uskov V N and Chernyshov M V 2014 Extreme shock wave systems in external aerodynamic problems. *Thermophysics and Aeromechanics* **21** (1) pp 15-31
[12] Silnikov M V and Chernyshov M V 2018 Supersonic flow at an overexpanded nozzle lip. *Shock Waves* **28** pp 765-784

Information system of remote sensing data analysis

A V Bakhtina[1], Y P Yuronen[1] and E A Yuronen[1,2]

[1]Reshetnev Siberian State University of Science and Technology, 31 Krasnoyarskii rabochii prospekt, Krasnoyarsk, 660037, Russia
[2]Siberian Federal University, 79 Svobodny Av., 660041 Krasnoyarsk, Russia

E-mail: anna-denisyuk@yandex.ru

Abstract. The information system is designed to solve the problems of analyzing remote sensing data and making decisions using non-parametric pattern recognition algorithms, the synthesis of which is based on kernel estimates of probability density. Initial information is compiled by training samples on the states of the investigated natural objects and their corresponding spectral data, which are presented in Microsoft Excel files. The application of the software provides the following functions: primary processing of remote sensing data to evaluate the main quantitative characteristics of the laws of distribution of random variables; testing of the hypothesis on distributions of multivariate random variables based on non-parametric pattern recognition algorithms; estimating the state of natural objects by spectral data; mapping of territories by state of natural objects based on spectral data of earth surface elements and their corresponding spatial coordinates. Results of the presented functional tasks are saved in the form of reports in graphic files and spreadsheets of Microsoft Excel. Software tools can be used in the study of various nature objects from remote sensing data.

1. Introduction

Efficient use of natural resources requires the development of information tools for their control. The most advanced means of natural resources research are remote sensing techniques involving aerospace vehicles. Therefore, the development of information tools for integrated analysis of remote sensing data is the basis for the creation of geographic information systems that ensure the speed of obtaining the required information and support for management decision-making. Their use makes it possible to significantly increase the accuracy and speed of solving decision-making tasks, which is especially relevant when developing the natural resources of the North and Siberia. At the same time, one of the main problems of decision making is testing statistical hypotheses about distributions of multivariate random variables, estimating the states of examined objects and their spatial mapping [1-4].

This paper proposes information tools for solving the above problems using modified non-parametric pattern recognition algorithms and the results of their application in the analysis of remote sensing data of forest areas.

2. Characteristics of software for estimation of natural objects states

Software functionality:

- Primary processing of remote sensing data providing assessment of basic quantitative characteristics of laws of random variables distribution (mathematical expectation and its reliable estimation, standard deviation, coefficients of asymmetry, kurtosis and antikurtosis).

- Testing hypotheses on distributions of multidimensional random variables using non-parametric kernel-type pattern recognition algorithms.
- Estimation of the state of natural objects by spectral data based on non-parametric pattern recognition algorithms.
- Spatial distribution of natural objects states according to remote sensing data.

The main source information for remote sensing data processing is a training sample $V = \left(x_1^i, \ldots, x_k^i, \delta(i), i = \overline{1,n}\right)$ composed of spectral data of $x^i = \left(x_1^i, \ldots, x_k^i\right)$ earth surface elements S^i and indications $\delta(i)$ about their belonging to natural object states (classes) Ω_j, $j = \overline{1,M}$.

Classes define the properties of Earth surface elements. Training and control samples are presented as Microsoft Office Excel spreadsheets. The number of classes (states of natural objects) is limited to 10. Number of used spectral channels (x features) - up to 15.

3. Estimation of natural object states

The basis of the software for estimating the states of natural objects is non-parametric pattern recognition algorithms, the synthesis of which uses kernel estimates of probability density estimates of the Rosenblatt-Parzen type [5, 6]. They are used to restore probability densities $p_j(x)$ in the area of definition, for example, j-th class by sample $V_j = \left(x_v^i, v = \overline{1,k}, i = \overline{1,n_j}\right)$ of volume n_j, which is composed of feature values of x_v, $v = \overline{1,k}$ objects of study. In this case, the non-parametric probability density estimate $p_j(x)$ is recorded as

$$\bar{p}_j(x) = \left(n_j \prod_{v=1}^{k} c_v\right)^{-1} \sum_{i=1}^{n_j} \prod_{v=1}^{k} \Phi\left(\frac{x_v - x_v^i}{c_v}\right). \qquad (1)$$

where $\Phi(u_v)$ – kernel functions satisfying the conditions:

$$\Phi(u_v) = \Phi(-u_v),\ 0 \le \Phi(u_v) < \infty,\ \int_{-\infty}^{\infty} \Phi(u_v) du_v = 1,\ \int_{-\infty}^{\infty} u_v^2\, \Phi(u_v) du_v = 1,$$

$$\int_{-\infty}^{\infty} u_v^m\, \Phi(u_v) du_v < \infty,\ 0 \le m < \infty,\ v = \overline{1,k}.$$

Let there be a training sample $V = \left(x^i, \delta(i), i = \overline{1,n}\right)$ for solving a multi-alternative pattern recognition problem, composed of the features values $x^i = \left(x_v^i, v = \overline{1,k}\right)$ of the objects of study and indications $\delta(i)$ about their belonging to one of the classes Ω_j, $j = \overline{1,M}$.

To evaluate the states of the object of research, we will use non-parametric algorithms for recognizing kernel-type images that meet the maximum likelihood criterion [7, 8]

$$\bar{m}(x):\quad x \in \Omega_j, \text{ if } \bar{p}_j(x) = \max_{t=\overline{1,M}} \bar{p}_t(x), \qquad (2)$$

where $\bar{p}_j(x)$ – non-parametric probability density estimate type (1).

Selection of optimal blur coefficients c_v, $v = \overline{1,k}$ of kernel functions in algorithm (2) is carried out on condition of minimum estimation of probability of classification error $\bar{\rho}$ in "sliding exam" mode according to training sample V. Optimizing a non-parametric solving rule (2) can be simplified by putting $c_\nu = c\bar{\sigma}_\nu(j)$, $\nu = \overline{1,k}$, where $\bar{\sigma}_\nu(j)$ – estimation of standard deviation of random variable x_ν in class Ω_j. This statement is obvious because the longer range of values of feature x_ν corresponds to the larger blur coefficient c_ν of kernel function $\Phi(u_\nu)$, $\nu = \overline{1,k}$. A similar approach is used in the construction of "fast" procedures for optimizing non-parametric estimates of kernel type probability density [9-15].

4. Assessment of natural object states using non-parametric classification algorithms with advantage gradations

When using a non-parametric algorithm (2), conditions of "reliable" and less "reliable" solutions arise. Under conditions $x = x'$ for "reliable" solutions $x' \in \Omega_j$ he value of the non-parametric probability density estimate $\bar{p}_j(x')$ should "significantly" exceed $\bar{p}_t(x') \neq 0$, $t \in J(x')$, where $J(x')$ – is a set of class numbers in which $\bar{p}_t(x')$ is different from zero at $t \neq j$. The concept of "significantly" gives rise to uncertainty in making decisions about the belonging of the situation x' to a particular class. To take into account the objectively arising problem, we introduce the threshold q, which we will compare with the ratio $\bar{q} = \bar{p}_j(x') / F(\bar{p}_t(x'), t \in J(x'))$, where $\bar{p}_j(x')$ is determined by statistics (1) at $x = x'$. For example, $F(\bar{p}_t(x'), t \in J(x'))$ can be defined by values:

$$\bar{p}_\lambda(x') = \max_{t \in J(x')} \bar{p}_t(x')$$

or

$$F(\bar{p}_t(x'), t \in J(x')) = \sum_{t \in J(x')} \bar{p}_t(x').$$

The threshold q is assigned by the researcher in accordance with the specifics of the problem being solved, thereby determining the "reliability" of the decisions taken. Higher $\bar{q}$ values compared to q determine a more "reliable" solution $x' \in \Omega_j$. This approach increases the number a priori of classes entered from M to $\bar{M}$, where $\bar{M}$ depends on the value of M and the number of thresholds r. With r=1 value $\bar{M} = 2M$.

Let us denote the new classes through $\Omega_j(q)$ and $\bar{\Omega}_j(q)$, which differ in the relations between the threshold q and the value $\bar{q}$ under conditions $x = x'$. Moreover, class $\Omega_j(q)$ corresponds to more "reliable" solutions $x' \in \Omega_j$, and $\bar{\Omega}_j(q)$ – less "reliable" in the conditions of the accepted threshold q. In this case, the non-parametric pattern recognition algorithm (2) with advantage gradations at $r = 1$ will be recorded as

$$\bar{\bar{m}}_1(x') : \begin{cases} x' \in \Omega_j(q), \text{if } \bar{q} \geq q \text{ } or \text{ } \bar{p}_j(x') > 0, \text{ and } \bar{p}_t(x') = 0, t = \overline{1,M}, t \neq j \\ x' \in \bar{\Omega}_j(q), \text{if } q > \bar{q} \geq 1, \ j = \overline{1,M} \ . \end{cases} \quad (3)$$

5. Application of an information system

The task of assessment of forest woodlands states according to remote sensing data obtained from Landsat 8 was considered. Each element of the earth's surface was characterized by six spectral features $x = (x_1, \ldots, x_6)$ in the wavelength range 450-2300 nanometers.

The volume of the training sample V was determined by the value n=1701. Source snapshot (figure 1a) was presented as a reference sample of volume n'=39644. The state of the natural site

Ω_j, $j=\overline{1,3}$ corresponded to damaged forests, coniferous and deciduous woodlands. Estimation of error probability of determination of forest woodland states using non-parametric algorithm (2) is determined by value $\overline{\rho}(\overline{c})$=0,035.

The results of using the proposed information tools in processing the initial image are given in figure 1.

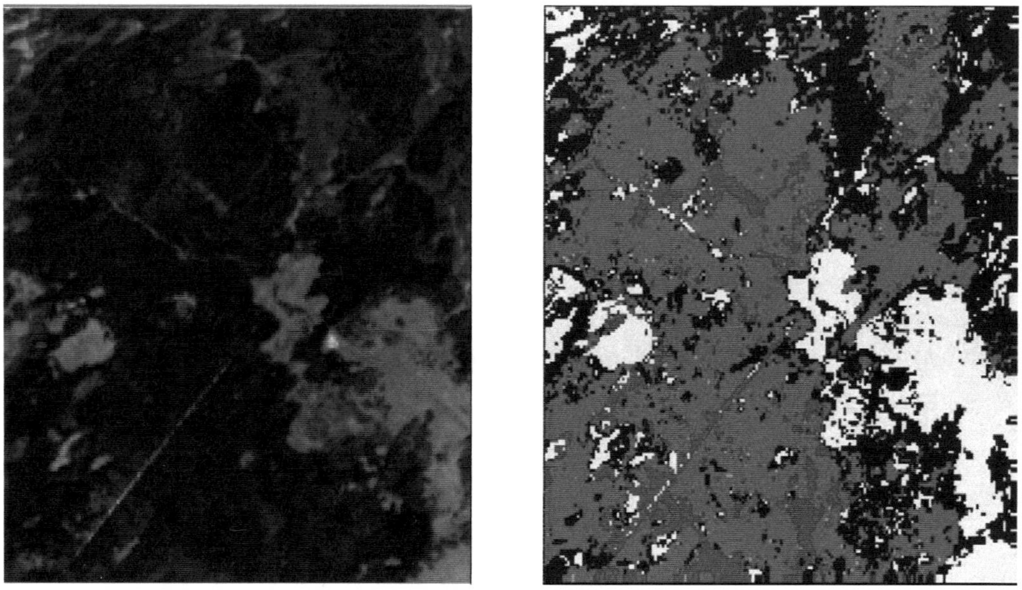

a) b)

Figure 1. Initial picture of the territory from the Landsat 8 (a) and the result of its processing using the developed information system (b).

In figure 1b sections of territories that correspond to damaged woodlands are marked in red. Green and blue colors correspond to areas with coniferous and deciduous trees. Fragments of territories that do not belong to the specified classes are represented in white.

6. Conclusion

The information system developed in the Delphi IDE provides an effective solution to the problems of estimating the natural objects states and their spatial distribution, as well as testing hypotheses about distributions of multidimensional random variables according to remote sensing data. They are based on non-parametric kernel-type pattern recognition algorithms, including classification methods with gradations of advantage. The representation of kernel function blur coefficients for non-parametric classification algorithms $\overline{m}(x)$ for each feature x_v, $v=\overline{1,k}$ as the product of the dimensionless factor c per standard deviation estimate x_v makes the optimization problem much easier $\overline{m}(x)$.

References

[1] Lapko A V and Lapko V A 2018 Nonparametric algorithms for estimating the states of natural objects *Optoelectronics, Instrumentation and Data Processing* **54** pp 451-456

[2] Lapko A V and Lapko V A 2019 A technique for testing hypotheses for distributions of multidimensional spectral data using a nonparametric pattern recognition algorithm *Computer Optics* **43(2)** pp 238-244

[3] Lapko A V, LapkoV A, Im S T, Tuboltsev V P and Avdeenok V A 2019 Nonparametric algorithm of identification of classes corresponding to single-mode fragments of the

probability density of multidimensional random variables *Optoelectronics, Instrumentation and Data Processing* **55** pp 230-236
[4] Lapko A V and Lapko V A 2019 Information system for natural objects state estimation according to remote sensing data based on nonparametric decision-making algorithms *Informatika i sistemy upravleniya* **59(1)** pp 106-117
[5] Parzen E 1962 On estimation of a probability density function and mode *Annals of Mathematical Statistics* **33(3)** pp 1065-1076
[6] Epanechnikov V A 1969 Non-parametric estimation of a multivariate probability density *Theory of Probability & Its Applications* **14(1)** pp 156-161
[7] Lapko A V and Lapko V A 2010 Analysis of asymptotic properties of nonparametric estimate of the equation of the separation surface in a two-alternative problem of pattern recognition *Optoelectronics, Instrumentation and Data Processing* **46(3)** pp 243-247
[8] Lapko A V, Lapko V A, Sokolov M I and Chentsov S V 2000 *Nonparametric classification systems* (Novosibirsk: Nauka)
[9] Lapko A V and Lapko V A 2018 Fast algorithm for choosing kernel function blur coefficients in a nonparametric probability density estimate *Measurement Techniques* **61** pp 540-545
[10] Lapko A V and Lapko V A 2019 Fast algorithm for choosing blur coefficients in multidimensional kernel probability density estimates *Measurement Techniques* **61** pp 979-986
[11] Lapko A V and Lapko V A 2019 Selection of the blur coefficient for probability density kernel estimates under conditions of large samples *Measurement Techniques* **62** pp 383-389
[12] Lapko A V and Lapko V A 2019 Methods for rapid selection of kernel function blur coefficients in a nonparametric pattern recognition algorithm *Measurement Techniques* **62** pp 300-306
[13] Lapko A V and Lapko V A 2019 Fast selection of blur coefficients in a multidimensional nonparametric pattern recognition algorithm *Measurement Techniques* **62** pp 665-672
[14] Lapko A V and Lapko V A 2019 Method of fast bandwidth selection in a nonparametric classifier corresponding to the a posteriori probability maximum criterion *Optoelectronics, Instrumentation and Data Processing* **55(6)** pp 597-605
[15] Lapko A V and Lapko V A 2020 Dependencies between histogram parameters and the kernel estimate of the probability density of a multidimensional random variable *Measurement Techniques* **62(11)** pp 945-952

Method for joint tracking by using magnetic field of the welding current during arc welding

V Ya Braverman and V V Bogdanov

Reshetnev Siberian State University of Science and Technology, 31 Krasnoyarskii rabochii prospekt, Krasnoyarsk, 660037, Russia

E-mail: braverman-vladimir@rambler.ru

Abstract. We investigated a possibility to determine electrode deviation from the joint during arc welding by using the strength of the magnetic field from the components of the welding current flowing through the welded parts. Amplitude of the vertical component of the resulting magnetic field is proportionate to the deviation of the electrode from the joint and the sign determines the direction of the deviation. The differential flux gate is utilized as a sensor for the electrode deviation from the joint.

1. Introduction

Automation of the technological welding processes has a number of challenges, one of which is the insufficient precision of automatic systems used to control position of the welding tools (electrode) at the joint of the welded parts. This may lead, first, to a possibility of a permanent defect and, second, a significant probability of producing joints of poor quality that may require rework and result in a longer technological process and an expansion of the production cycle. Additionally, control systems often function as a part of an expensive equipment set selected for the unique nature and the cost of the welded items and therefore, the issue of quality may create substantial economic consequences. Many control systems have an electrode position sensor that is firmly connected to the welding tool and is located at a certain distance from the welding site. Hence, the sensor measures the deviation from the sensor and not from the tool. This creates a methodical error of aligning the electrode with the joint. Additionally, high level of electromagnetic noise is present under the conditions of the welding process and it negatively impacts functionality of the joint sensor of the control system [1-3]. The challenge of delivering required precision for the automatic control systems guiding the placement of the welding tool has had limited solutions up to now and this is the subject of our current work.

2. Problem Statement

The method is based on the identification of the magnetic field of the welding current (I_w) and the position of the electrode at the joint. The main principle of this method is based on the fact that when the electrode deviates from the joint, there is a redistribution of the welding current components and the magnetic fields caused by them [4], [5]. With the help of the current collector, the welding current I_w is split into two components I_1 и I_2 (figure 1). If the electrode is positioned precisely above the joint, then the strength of the magnetic fields H_1 и H_2 (and their vertical components H_{1v} и H_{2v}) is of equal value but of the opposite direction (figure 1.a). The resulting magnetic field in this case is determined by the electrode current I_w, and the vector of the magnetic-field strength is located in horizontal plane.

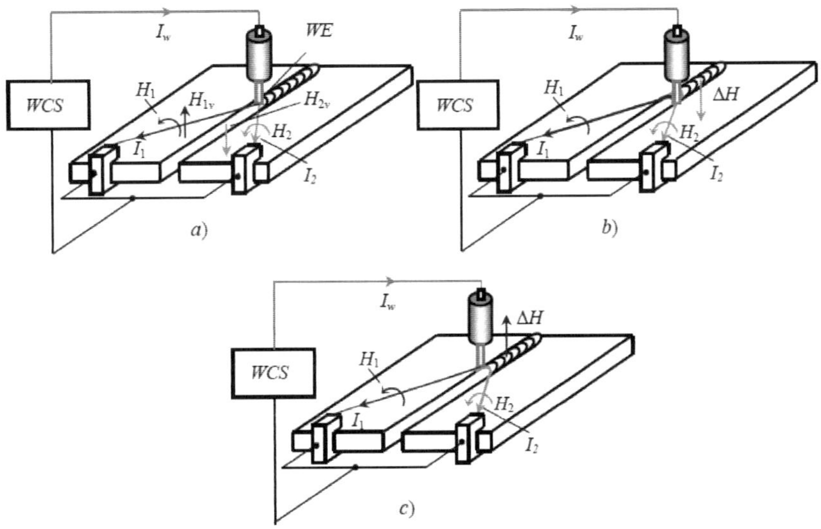

Figure 1. Formation of the vertical component of the magnetic field with the strength ΔH above the welded section from the current components I1 and I2 when the electrode shifts.

When the electrode deviates from the joint (figure 2, *b*), the welded segment appears along component I_1 of the welding current I_w from the electrode to the current collector and the strength of the magnetic field caused by the current I_1 will decrease as I_1 decreases when the current passes the welded segment (increasing current resistance I_1). The resulting magnetic field is defined by the vertical component of the strength ΔH and the horizontal component H_w from the current I_w. When the electrode deviates in the opposite direction, the welded segment appears along the current I_2 and the strength of the magnetic field caused by current I_2 will decrease as I_2 decreases when the current passes the welded segment (increasing current resistance I_2). The resulting magnetic field is defined by the vertical component of the strength with the opposite direction compared to the prior example and the horizontal component H_w. Therefore, the vertical component of the magnetic field carries information regarding the electrode's position at the joint.

3. Experimental Modeling

He physical model [6] provides experimental proof for changes in the electromagnetic conditions depending on the position of the electrode at the joint. Electrolyte solution is used as a conducting matter (material of the welded parts). Dielectric plate is used to imitate the joint. Voltage source, current supply, Electrolyte solution, electrode dipped in the solution at the edge of the welding seam and the joint create a model of the electric chain of the welding current. A family of equipotential curves can be registered on the solution surface with the help of a probe and an indicator. For this purpose, a set of points with the same potential can be found when the probe is moved in the direction across the joint with a given step along the length coordinate. Then the mesh of power lines of the current can be built over the family of the equipotential curves. As the mesh is generated, one can rely on the fact that the power lines in any point of the field must be normal in relation to the equipotential curves (figure 2). It is clear that when the electrode deviates from the joint, the symmetry of the current distribution is disturbed (figure 2,*b*). This means that the resulting magnetic field is also different from the field when there is no misalignment of the electrode and the joint (figure 2,*a*).

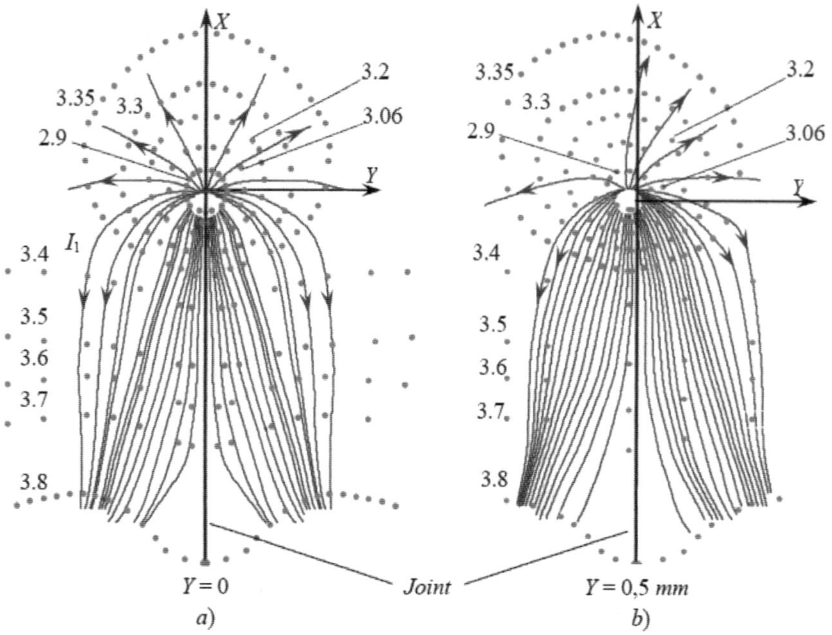

Figure 2. Results of physical modeling of the current distributions I1 и I2 (values mark the potential).

4. Analytic researches

The dependence of the magnetic field power on the magnitude of the electrode deviation from the joint is determined analytically by using a three-dimensional model, reflecting the process of the arc welding with respect to the "electromagnetic" properties of the near-electrode environment, limited to the surface of the welded parts with coordinates derived by analyzing the two-dimensional model. This dependency can be represented as follows:

$$H_z = F(x, y, h_\partial, I, \Delta), \quad (1)$$

where H_z – vertical component of the magnetic field; x – length coordinate of the joint; y – cross coordinate of the surface of the welded parts; h_∂ – height of the sensor position; Δ - excess weld; I – welding current. The modeling consists of the joint solution of the Maxwell equations:

$$\begin{aligned} rot\vec{H} &= \vec{j} + \frac{\partial D}{\partial t} \\ rot\vec{E} &= -\frac{\partial \vec{B}}{\partial t}, \\ div\vec{D} &= \rho, \\ div\vec{B} &= 0, \end{aligned} \quad (2)$$

where E- power of the electric field, D – displacement field (electric current), H- voltage of the magnetic field, j- density of the current, ρ- volume density of the electric charge, B – magnetic induction.

The system of Maxwell equations is supplemented by material equations that connect vectors E, D, H and B with values describing electric and magnetic properties of the environment:

$$\begin{aligned} \vec{D} &= \varepsilon_0 \varepsilon \vec{E}, \\ \vec{B} &= \mu_0 \mu \vec{H}, \\ \vec{j} &= \gamma \vec{E}, \end{aligned} \quad (3)$$

where ε- relative dielectric penetration, μ – relative magnetic penetration, γ- electrolytic conductivity, ε_0- electric constant, μ_0 – magnetic constant.

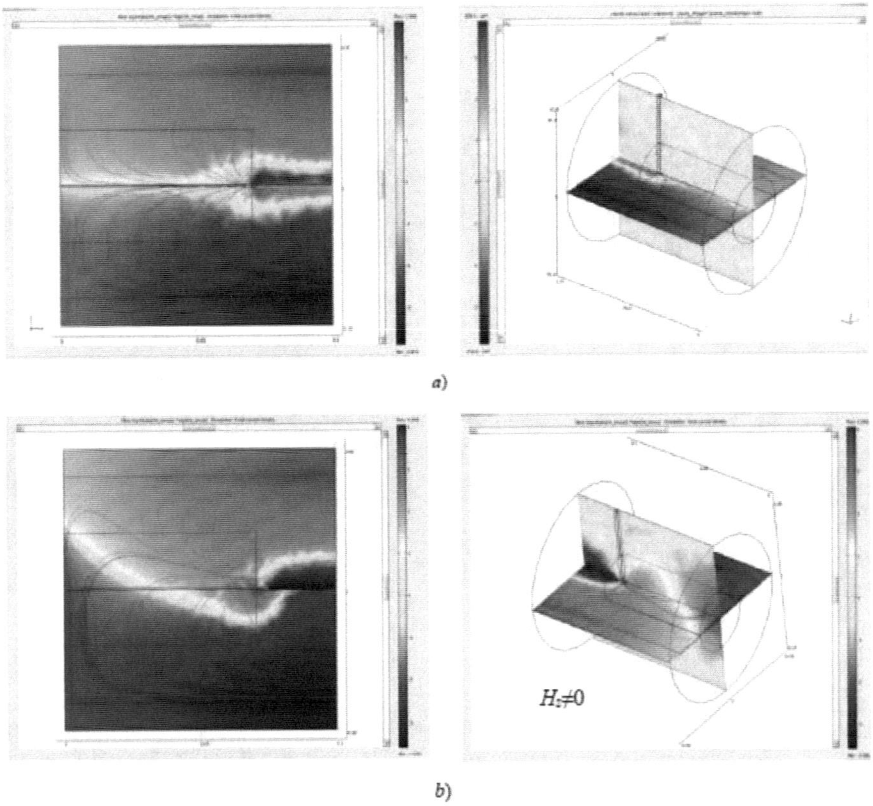

Figure 3. Distribution of magnetic fields: a – no shift of the electrode; b – the electrode shifted relative to the joint by 0.5 mm

The current enters the circuit through the electrode and is collected from the side of the unwelded part of the joint. The parts are surrounded by the cylindrical layer of air, in the vicinity of which magnetic field needs to be determined with the varying y-coordinates of the electrode. The environment under observation is considered isotropic linear with dielectric and magnetic parameters ε and μ respectively expressed as real numbers because of the low frequency of the observable processes. Borderline conditions: for the electric component: voltage up to 1,000 *A* is applied to the upper edge of the element "arc"; side edges from the cut surface – zero potential; other edges – electric insulation. Calculations are performed with «*Comsol Multiphisics*» software. We modeled two parameters – constant current in the conductive environment and the electromagnetic field of the constant current. We solved the equations with the finite element method for splitting into physical processes and coordinates on the mesh with variable intervals. Calculations with the variable shift of the arc, welding current and excess weld are performed for materials of aluminum alloys (μ=1) and steel (μ=4000) with the local thermal equilibrium in the area of the thermal impact of the welding process.

Below are the modeling results for welding parts made of AMg-6 alloy with the thickness δ = 10 mm and I_w = 300 *A* and the welding velocity S_w=0.2 *sm/sec*. It is clear that in the absence of the electrode shift distribution of the components I_1 и I_2 of the welding current is symmetrical relative to the joint line (figure 3, *a*). When the electrode shifts away from the joint, the symmetry is disrupted (figure 3, *b*) and a portion of the current of the welded part with the electrode deviation, goes through the welded segment. The greater the electrode deviation, the greater is the symmetry disruption which aligns with the results of the field modeling. As the equilibrium of the currents I_1 и I_2 is disrupted, a vertical component of the magnetic field H_z emerges with the value that depends on x- and y-coordinates (figure 3). The right side of the figure shows the distribution of H_z. The modeling results reveal that the deviation of the electrode from the joint transforms (among other things) into the change of the vertical component of the magnetic field. The equation for H_z may be presented as follows:

$$H_z \approx \frac{1}{4\pi r_0} \cdot \frac{I \cdot y}{\sqrt{r_0^2 + y^2}}, \tag{4}$$

where r_0 – average distance from the field source to the measurement point; I – average rate of the current. Calculations show that H_z is essentially proportionate to the deviation of the electrode form the joint in a wide range (figure 4).

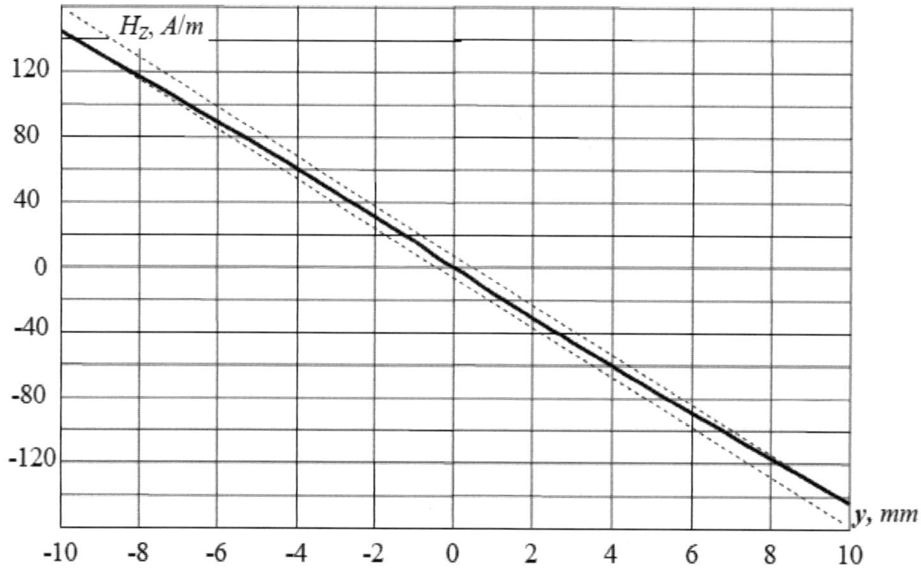

Figure 4. Dependency of H_z on y

Results of our work demonstrate that it is possible to determine electrode position by using the rate and direction of the vertical component of the magnetic field of the currents. At the same time, if you position the sensor at a certain distance from the electrode, this will not result in a methodical error since H_z appears when the electrode deviates from the joint.

5. The system for joint tracking

Figure 5 is a functional diagram of the system for automatic joint tracking with flux gate as a sensor for the position of the electrode relative to the joint of the welded parts. The system includes the following: source of the welding current *WCS;* flux gate sensor *S* for the position of the electrode relative to the joint; control system *CS*; electric drive *ED*, moving the welding head in the direction perpendicular to the joint line. The control system includes amplification-conversion elements. The sensor (for example, differential flux gate) is firmly connected with the welding electrode. The flux gates's axis of sensitivity is oriented in the vertical plane perpendicular to the welding surfaces. When the sensor is positioned this way, it becomes insensitive to the magnetic field of the welding current I_w.

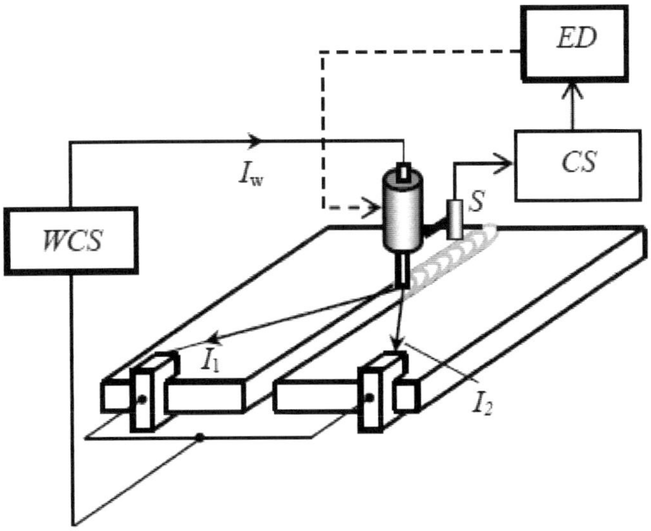

Figure 5. Structural diagram of the joint tracking device

When the electrode is positioned at the joint (figure. 1, *a*), the voltage projection of the magnetic field from the welding current above the welded segment on the flux gate sensitivity axis is equal to zero and there is no exit signal from the flux gate. When the electrode deviates from the joint (figure 1, *b*), the component of the welding current I_1 flows through the welded segment of the joint and voltage projection of the magnetic field on the flux gate sensitivity axis indicates the magnitude and the direction of the deviation of the electrode from the joint. Under the influence of the voltage in the measuring coil of the flux gate electromotive force is identified. The flux gate exit signal is amplified and the voltage of the constant current forms at the *CS* exit. This voltage is proportionate to the magnitude of the electrode's shift form the joint. The sign of the signal determines the direction of the deviation. The voltage enters the electric drive *ED*, which eliminates misalignment of the electrode's position and the joint. The device was tested under industrial conditions. The error of aligning the electrode and the joint does not exceed 0.4 mm.

5. Conclusions

1. We are proposing a method to determine position of the electrode at the joint based on the identification of the magnetic field currents and the electrode coordinates.

2. This method allows us to control the actual location of the electrode and correct the location with the help of an automatic control system without methodical errors.

3. We increased interference resistance of the control system. Unlike in other existing systems where the welding current is a significant source of interference, in our system, the welding current becomes the information source.

References

[1] Reichmann A, Leffler D and Bartel R 2004 State of the art of a FEP control unit of beam scanning for electron beam equipment *6th International Conference on Beam Technology*, (Halle) pp 102-107

[2] Nikiforov D D and Lifshitz M L 1985 *TV Sets as Welding Elements of Nuclear Equipment* (Moscow, Energoatomizdat) p 126

[3] Gladkov E A 2006 *Process and Equipment Control during Welding* (Moscow, Publishing Center «Akademia») p 432

[4] Braverman V Y, Laptyonok V D and Belozertsev V S 1982 Fluxgate Sensor for Joint Tracking *Copyright certificate No 941056 USSR, international classification index: B23K9/10,* available at: https://patents.su/4-941056-ferrozondovyjj-datchik-dlya-slezheniya-za-stykom-svarivaemykh-detalejj.html

IOP Conf. Series: Materials Science and Engineering 1230 (2022) 012003 doi:10.1088/1757-899X/1230/1/012003

[5] Braverman, V Y, Laptyonok V D and Belozertsev V S 1982 Method for Measuring Deviations of the Electrode from the Seam. *Copyright certificate No 941057 USSR, international classification index: B23K9/10*, available at: https://patents.su/4-941057-sposob-izmereniya-velichiny-otkloneniya-ehlektroda-ot-styka.html
[6] Belozertsev V S 2010 *Increasing Precision of the Control System for the Welding Tools in Technological Processes of the Arc Welding* (Krasnoyarsk, Reshetnev SibGAU) p 194

Implementation of a combined algorithm designed for increasing the reliability of information systems

S S Bychkov [1], **A M Popov** [2] **and A K Gorbacheva**[2]

[1] Krasnoyarsk State Government Agency "Information Technology Center of Krasnoyarsk Krai", 2 Robespierre St., Krasnoyarsk, 660021, Russia
[2] Reshetnev Siberian State University of Science and Technology, 31 Krasnoyarskii rabochii prospekt, Krasnoyarsk, 660037, Russia

E-mail: ampopov1959@gmail.com

Abstract. This paper considers the results of experimental studies of the previously presented algorithm designed for increasing the reliability level of information systems. The data on the organization and conduct of the studies is provided to clarify their meaning. Within the framework of the study conducted, the obtained experimental data of simulation modeling and the functioning data of existing information systems are compared. The hypothesis of homogeneity of the structure of information systems has been formed, thereby reconfiguring the presented algorithm into a model for the analysis of information systems. The results presented can be used in further research activities. The obtained data on the opportunity to predict the functioning of information systems can be used for economic planning.

1. Introduction

Disruption in the operation of the company in modern economic conditions leads to significant economic losses and reputational issues. In this regard, increasing attention is being paid to improving the reliability of production elements that have a direct impact on the organization's business process, while the level of process automation is steadily growing. Thus, increasingly greater attention is being paid to ensuring the efficiency of automation systems (AS), as well as control information systems (IS).

As a result, the efficiency (capacity to carry out a part of the production process correctly) of IS and AS is a parameter that affects the functioning of the company directly and must be given due particular consideration and analysis (no costs to restore IS/AS and correction of internal errors, fulfillment of obligations by hiring third parties to enforce contractual obligations) [6].

The authors have previously presented the paper describing the notion of the application options for the combined algorithm designed for increasing the level reliability of information systems [1,8] (Algorithm). While designing the Algorithm, the following mathematical methods have been selected:

- methods of calculating the reliability of the information system index, in particular, the dynamic method (dynamic methods work on the condition that an information system is being operated, thus indicating effectively the level of reliability of the IS);
- mathematical modeling of IS/AS (Mathematical modeling is used to predict the level of reliability of the information system at a certain point) [2,4];
- methods of decreasing mean time between failures (MTBF) and decreasing the failure rate of the information system (these methods are applied to tackle the identified problems in IS/AS).

Aiming to test the operability and correctness of the Algorithm, an experiment has been set up on the model of IS. The results of the experiment have been compared with the results of the existing IS. Based on the analysis of the data obtained, the conclusion has been formulated: within the framework of the assumptions and limitations coming from the use of the model, these ISs can be represented as a multiphase multichannel open-loop queuing system that is defined by the following features:

- input data stream;
- processing and combining data modules;
- data verification queues;
- multichannel service devices (control modules, processing module, data transmission modules);
- output data streams.

Exactly the same conclusion can be made as regards the majority of automatic control systems, where the data goes through the stages of modules. In this case, the data is being processed and changed, achieving the stage of output data, which can be used in managerial decision-making. In addition, it is also possible to use this approach for systems with dynamics of change, including that which is based on change management algorithms [9].

One of the main tasks of the experiment was to obtain data on the probable location (software/hardware module) and time (from the start of the computing process) of the occurrence of an error in the mathematical model of the IS. For the experiment, a simulator has been used. The decision of constructing a simulation model was made due to the availability of data on the functioning of the existing IS (the period for collecting statistical data is 5 years), as shown in table 1. During this period of time statistical duty has recorded all errors that occur in the IS, and the measures to eliminate them have been taken as well. For simulation, functional simulation software, AnyLogic produced by The AnyLogic Company, was used.

The necessity to apply the simulation in the presented experiment is based on the following factors:

- no need to use large amounts of computing power to run IS;
- possibility of altering the key parameters of the IS without exposing the IS to crash risk;
- possibility to select and combine the parameters of the simulation model that will allow to obtain accurate data based on the collected statistics (congruence of simulation results and actual statistical surveys) due to the proper selection of control coefficients;
- possibility of making predictions about the occurrence of failures in the IS based on the current simulation model;
- assessing the correlation of parameters of the simulation model, as well as reaching the adequate conclusions and recommendations concerning the real IS, including the use of time-to-failure methods.

Therefore, the use of the simulation model provides us with ample opportunities for carrying out a detailed analysis of the IS by setting up many experiments, without the risk of economic loss. There is an exceptional opportunity to make changes to the key parameters of the IS and its components [5,7].

Description of the studied IS[1].

Table 1. Input data for research.

No	IS	Parameter	Description	Quantity
1	IS 1	IS module (unique type of module)	- receiving information - initial verification of received information - recording of received information	6

[1] To preserve the confidentiality of the information, the IS names have been changed, no unique names of the modules as well as no unique features are included in the description, and the IS structure is simplified as well, which allows preserving the accuracy of the data during the experiment, but not allowing to discredit the specific IS.

		Data processing module (unique type of module)	-data alteration (the following types of modules are known: processing (conversion of information that involves minor changes), changes(getting totally different information comparing to initially received), storage and generation (getting new data without preliminary processing of received data, where the former is related to the latter))	6
		Data type (unique data category)	The number of types of incoming and outcoming data streams that cannot be merged and that are processed in the separate IS modules	8
		Time of functioning of IS (days)	The period of time during which the observation of the real IS has been conducted	2000
2	IS 2	IS module (unique type of module)	- receiving information - initial verification of received information - recording of received information	25[2]
		Data processing module (unique type of module)	The IS function module, which performs certain operations on the data alteration (the following types of modules are known: changes (getting altered (moderated) information comparing to initially received), storage)	27
		Data type (unique data category)	The number of types of incoming and outcoming data streams that cannot be merged and that are processed in the separate IS modules	25
		Time of functioning of IS (days)	The period of time during which the observation of the real IS has been conducted	2000
3	IS 3	IS module (unique type of module)	- receiving information - initial verification of received information - recording of received information	4
		Data processing module (unique type of module)	- data alteration (the following types of modules are known: processing (getting totally different information comparing to initially received), storage and generation (getting new data without preliminary processing of received data, where the former is related to the latter))	6
		Data type (unique data category)	The number of types of incoming and outcoming data streams that cannot be merged and that are processed in the separate IS modules	5
		Time of functioning of IS (days)	The period of time during which the observation of the real IS has been conducted	2000

Formal description.
According to the data presented, the IS models should consist of the following elements:
- simulation of data (of various types) entry;
- simulation of IS modules functioning;
- simulation of data processing modules;

[2] These modules receive the same type of information from different sources and send it to the same type of modules for processing. The result of processing will be identical regardless of the module. This IS is a striking example that indicates the option for reducing the simulation to a linear structure.

– simulation of output data verification modules functioning;
– simulation of data output.

Logic diagrams for the IS models 1, 2 and 3 are represented in figure 1, 2 and 3.

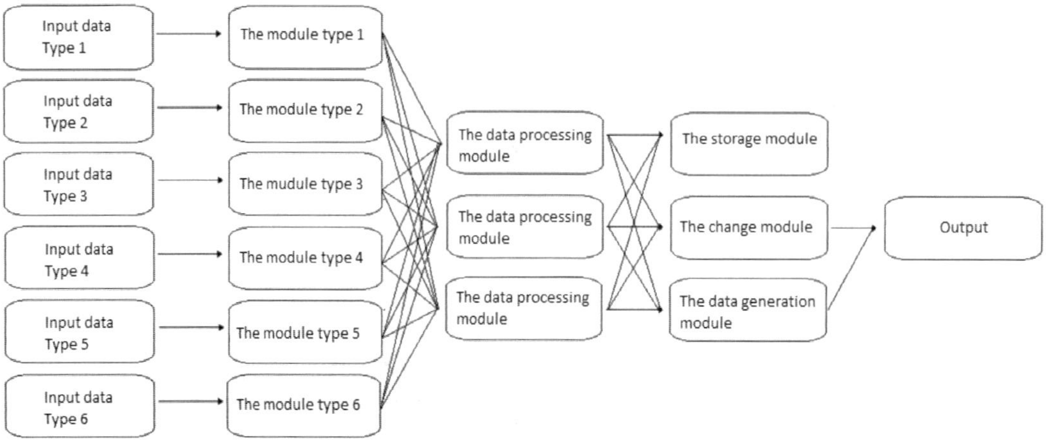

Figure 1. Logic scheme of the simulation model for the IS 1.

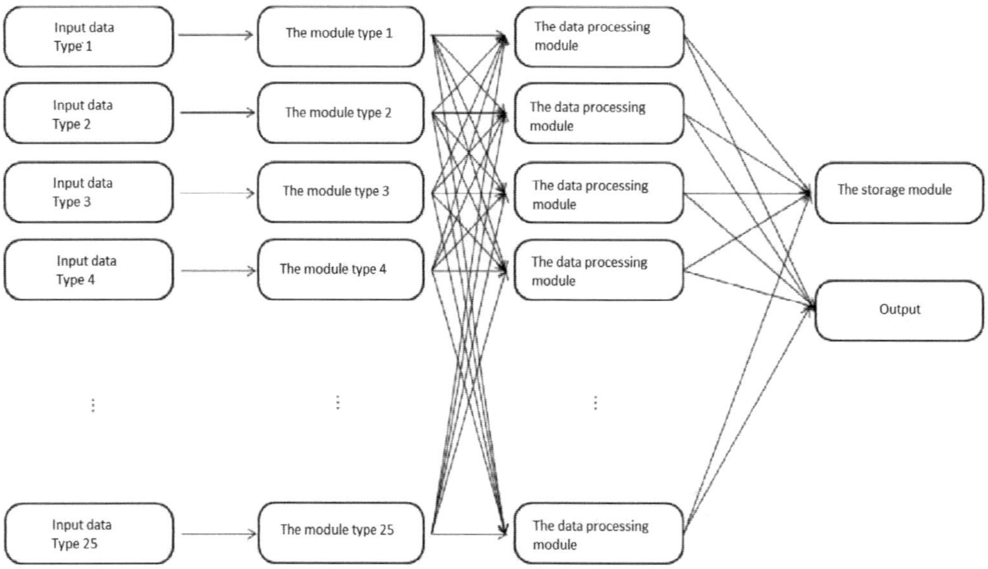

Figure 2. Logic scheme of the simulation model for the IS 2.

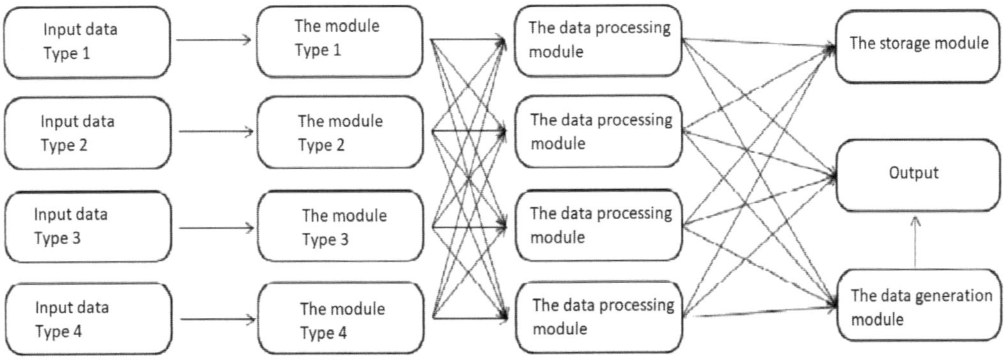

Figure 3. Logic scheme of the simulation model for the IS 3.

n order to analyze the operation of the IS the following IS-model experiment has been set up:

Several types of data are fed through the input of the IS model. The intervals between occurrence of data of the same type are random. Each type of data enters the corresponding IS module, after which it is exposed to the processing, the result of which is arbitrary (data can be received, modified and fed through the output of the IS, as well as can cause the generation of new data in the IS), and it cannot take more than a predetermined period time that have been laid down for processing the received data. After processing, the data are verified in one of the emission modules. The data verification time is arbitrary. After verification, part of the data can be returned for revision, the rest of the data comes out of the IS model.

As a result of the IS operation, errors can occur in the modules. The number of errors in the modules depends on the workload (the amount of processed data at the same time) of the modules, while an error in an arbitrary module leads to downtime of the whole model of the IS. The IS downtime is contingent on the restoration process of the IS and it is clearly fixed.

As can be noted from the represented diagram of the simulation IS model, possible errors in the IS operation will occur in modules that are aligned with the reception, processing, storage, generation of new data and data alteration, which is confirmed by the statistical data on the operation of the existing IS.

In this regard, errors that caused failures in the IS operation were installed (artificially programmed) in the IS models. Programmed failures are of probabilistic (conditionally random) nature, which is related to the frequency of operation of a particular IS module. Moreover, each failure is recorded leading to the decreased probability of the next one. Thus, the simulation of failures caused by disruption of the software and hardware parts of the IS is carried out, and therefore the human factor impact is eliminated completely.

The refusal to consider errors caused by the human factor is motivated by the lack of objective ways how to preclude these errors. Moreover, data of related to this type of errors were excluded from the statistical data compilation.

2. Results from the analysis

The results of this experiment are:

- The operation stage of the IS was simulated, which can be compared with real data, the weights are adjusted, the simulation results are close to real ones.
- The predictions about the timing and location of failure occurrence in the IS is made.

The diagram represented in figure 4 shows the deviation of the operating results of the simulation model from the data gained from the statistical compilation based on the functioning of the existing IS. The unit of time is 1 calendar day. The diagrams are presented for three simulated ISs. As can be seen from the diagrams represented, the deviation in the estimated dates of the information system is not more than 10 calendar days, which is not significant within the long-term and medium-term planning (for a period of 3 or more years). The variability of deviation for IS1 is 28.51, for IS2 it is 34.27, and 39.84 for IS3.

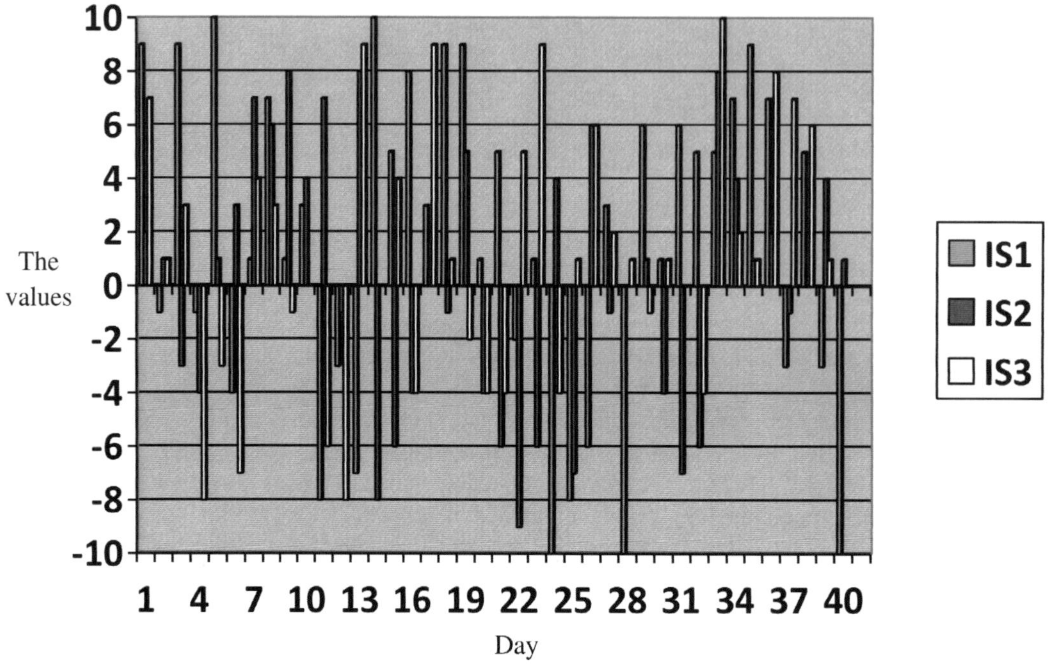

Figure 4. Deviations in the values obtained by IS models.

3. Conclusion

As a result of this research the following conclusions can be drawn:

- The simulation model allows us to carry out an analysis with high accuracy, pending the formalization of the model and the application of assumptions.
- A hypothesis has been formed: all ISs (including those that include hardware, such as ACS, APCS, etc.) can be represented as a queuing system. The confirmation of the hypothesis will simplify the construction of a unified model of the IS, the use of which will make it possible to simulate an arbitrary IS.
- The results obtained, subject to adjustments, can be used in planning operational indicators.
- The Algorithm can be applied in ensuring information security of the IS as related to identifying potential location for errors and zero-day vulnerabilities [3].
- The model can be used to design complex information security systems for automated and information parts of strategic and critical information infrastructure objects.

References

[1] Popov A, Zolotarev V and Bychkov S 2014 Improving the reliability of information systems *Siberian Journal of Science and Technology* **3** 55

[2] Azarnova T, Asnina N, Proskurin D and Polukhin P 2017 The formation of the Bayesian network structure of the process of testing the reliability of information systems *Bulletin of VSTU* **6**

[3] Mitrokhin V and Ringenblum P 2014 Assessing the impact of information security threats on the accessibility of the telecommunications network *TUSUR reports* **2** 32

[4] Litvyak R, Vorobyov S and Katsupeev A 2016 The model for studying the reliability of a high-availability information system based on architecture with redundant servers *University news. North Caucasus region. Series: Engineering* **2** 190
[5] Lvov A, Svetlov M and Martynov P 2014 Improving the information reliability of digital systems with qam/cofdm modulation *Izv. Saratov Univ. (N.S.), Ser. Math. Mech. Inform.* **4-1**
[6] Dorofeev A and Markov A 2015 Planning for business continuity and recovery *Issues of cybersecurity* **3** 11
[7] Melnik E and Klimenko A 2017 A comprehensive method for organizing the management of a fault-tolerant information-management system based on multi-agent interaction *Izvestiya TulGU. Technical Science* **9**
[8] Popov A, Zolotarev V and Bychkov S Implementation of a combined algorithm designed to increase the reliability of information systems: simulation modeling *IOP Conf. Series-Materials Science and Engineering* **155 (1)** 012010 (doi:10.1088/1757-899X/155/1/012010)
[9] Styugin M, Zolotarev V and Parotkin N 2019 Open information systems protection by its continuous change *Proc. 2019 RusAutoCon* 8867600 (doi: 10.1109/RUSAUTOCON.2019.8867600)

IOP Conf. Series: Materials Science and Engineering 1230 (2022) 012005 doi:10.1088/1757-899X/1230/1/012005

Using microwave irradiation to increase permeability of Siberian larch wood

S G Eliseev, V N Ermolin and S S Stupnikov

Reshetnev Siberian State University of Science and Technology, 31 Krasnoyarskii rabochii prospekt, Krasnoyarsk, 660037, Russia

E-mail: serezha-stupnikov@mail.ru

Abstract. The paper presents the research results of the microwave radiation effect on the Siberian larch heartwood. The collected experimental data show that microwave treatment increases the larch wood permeability not only longitudinal but also transversely the fibres. The most significant increase of the transversely fibre wood permeability is found in the tangential direction.

1. Introduction

In the last several decades, the interest for wood as an environmentally friendly renewable natural resource has been consistently growing. This is facilitated both by the unique wood properties and the growing welfare level of the developing countries ensuring the increasing consumption of natural wood as a more expensive and prestigious material.

The range of wood on the global market is represented by a broad variety of different wood species. Some wood species are sought for more than others due to their attractive colour, texture or good physical and mechanical properties. This creates the deficit of supply and, therefore, an increase in prices.

One of the ways to satisfy the demand for wood with the sought-for properties is a modification of less precious species growing in the northern latitudes, not remarkable for their high physical and mechanical properties, bright colour or distinctive texture.

One of the promising species for wood modification is Siberian larch wood (Larix sibirica) widely spread in the Siberian region [1]. Larch wood is known for its good physical and mechanical properties equal to those of such popular species as oak, ash, or beech [2] but has a pale colour and plain texture which can be successfully corrected with the modification methods depending on the purpose.

The majority of the wood modification methods are based on the treatment of wood with a modifying solution under excessive pressure. However, the larch wood treatment is almost impossible, as the major part of the trunk volume consists of the heartwood with low liquid and gas permeability [3].

The permeability-reducing mechanism in the softwood core formation process has been studied by numerous researchers [3-5]. In those studies, the low permeability of the heartwood is explained by the changed position of the toruses in the vertical tracheid pores and the deposition of the extractive substances in the marginal zones of the radial tracheid pores.

Increasing permeability of wood is one of today's important scientific and technological tasks. By the present time, a number of wood material permeability-increasing methods have been developed, including injection, alternating pressure treatment, wood-staining fungi implantation etc. [4-7]. The listed methods have their drawbacks, such as disintegration of wood structure, low energy efficiency, or technical complicatedness when applied in industrial conditions etc.

In recent time, scientists from all over the world are actively working on the methods for increasing the permeability of wood by means of microwave (MW) irradiation radiation treatment. For instance, the works [8-13] remark the high efficiency of (MW) radiation treatment of wood to increase its permeability with an acceptable reduction of its mechanic properties. The authors of [10] have found that the MW treatment of Chinese fir wood increases the treating solution absorption rate to 308% compared to the untreated reference sample; according to [11], the absorption rate increases to 156%. In the research [12], the MW radiation influence on the beech wood increased its longitudinal permeability by 28%, and in some cases — up to 156%.

In the quoted works [8-12], the MW radiation-induced permeability-increasing mechanism is explained, first of all, with the effect of heating and evaporation of the water contained in wood, producing steam-gas pressure. As a result, the cell walls of the wood elements are partially destroyed, softening the material and enforcing the movement of the extractive substances.

Noticeably, in the majority of the studied materials, the authors mostly focus on improving the longitudinal permeability. However, the determining factor for treatment of elements over one meter long is the transverse permeability of wood [3].

With all the collected data of the MW radiation treatment potential, it was decided to study the possibility of increasing the Siberian larch (Larix sibirica) heartwood permeability.

2. Experimental and Results

For the research, from a part of the 110-year-old larch tree log, 1.2 meters away from the butt, a block with the diameter of 40 cm was cut. From the block core, along the fibres, some samples of 50x50 and 150 mm with the similarly oriented growth rings were produced. The samples were divided into three groups of 10 pieces each. The first two groups were treated with MW radiation; the third one was the reference group.

To simulate the industrial batch amount, the longitudinal permeability of the samples was blocked. For this purpose, the sample ends were coated with a layer of silicon sealant and covered with plastic caps with rubber seals inside. Before placement into the MW chamber, the opposite caps were tightened with a heat resistant plastic clamp strap (figure 1).

The SHF treatment of the samples was carried out in the microwave chamber at the frequency of 2.45 GHz with a fixed power of 0.9 kW. At the moment of treatment, the humidity of samples was 55%. The duration and method of treatment were selected based on the published research data [10-12]: Group 1: MW treatment of 5.5 min; Group 2: MW treatment of 4 min; Group 3: untreated reference samples.

After that, the MW treated and untreated samples were thoroughly waterproofed (to exclude the end permeability) with PF-115 and silicon sealant applied in multiple layers. Then, the samples were treated in an autoclave under the pressure of 0.5 MPa for one hour. The treatment solution was the water solution of $FeSO_4$ which allows observing the treatment infiltration due to its wood colouring properties.

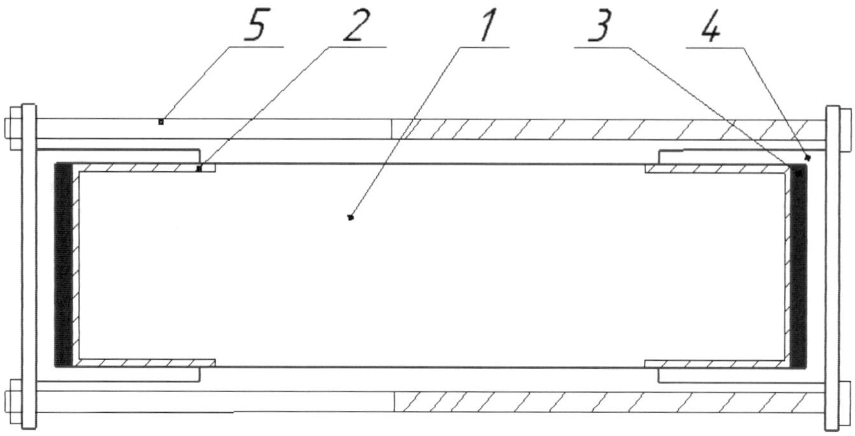

Figure 1. Preparation of samples for MW treatment (1. Sample; 2. Silicon sealant; 3. Rubber seal 4. Plastic cap; 5. Plastic bolt.

As a result of the studies, the MW treated samples were discovered to feature a larger treatment solution infiltration depth in the radial direction compared to the reference samples (figure 2). In the given conditions, the samples from Group 1 show over a 1.5-time increase in the treatment infiltration front depth progress, while the infiltration value of Group 2 exceeds that of the reference samples by 6% only. The infiltration depth of the samples from Group 1 (exposed to longer MW treatment) exceeded that of the samples from Group 2 by almost 60%.

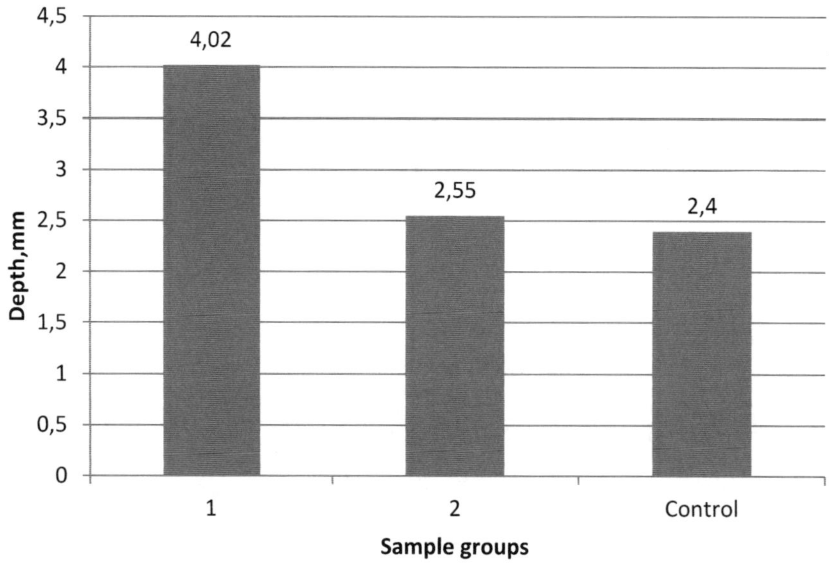

Figure 2. Treatment infiltration depth progress in the radial direction.

For the tangential infiltration (figure 3), the MW treated samples from both Group 1 and 2 show an almost double increase in the progress of the treatment infiltration depth. However, the longer MW treatment of the Group 1 samples did not induce a dramatic change in the treatment infiltration depth compared to those of Group 2. The difference between the samples of Groups 1 and 2 lies within the experiment accuracy limits.

Generally, at the maximum lasting exposure (Group 1), among the structural directions, the greatest permeability was observed in the radial direction with the treatment solution infiltration depth being

12% higher than that of the tangential direction. At the same time, tangential treatment infiltration depth progress value of the samples of Group 2 (shorter MW exposure) exceeded that of the radial infiltration depth progress by almost 1.4 times.

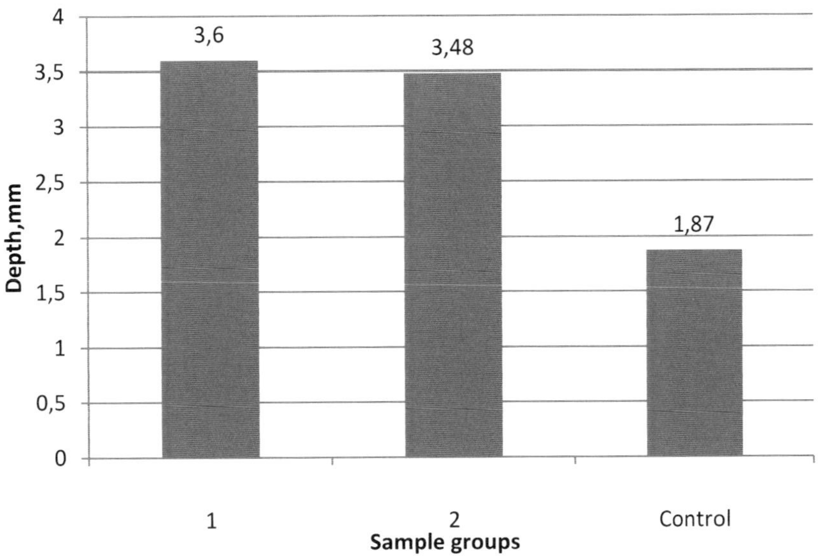

Figure 3. Treatment front depth progress in the tangential direction.

3. Conclusion

The collected experimental data showed that the MW treatment of the Siberian larch wood facilitates the increase of the wood transverse permeability. After MW treatment of the larch wood, the treatment solution infiltration depth progress was found to increase by 1.5 times in the radial direction and by 2 times in the tangential direction. The collected data prove the high efficiency of the MW radiation treatment as a way of increasing transverse permeability of the Siberian larch heartwood.

References

[1] Utkin A I, Zamolodchikov D G and Chestnykh O V 2003 *Trees of the Boreal Zone* 1 33
[2] Ugolev B N 2007 *Wood Studies and Timber* **1** 12
[3] Kharuk E V 1974 *Wood and wood materials* **1** 84
[4] Bazhenov V A, Kharuk, E V 1952 *Research in the field of wood and wood materials* **1** 62
[5] Moskaleva V E, Goncharova *Research in the field of wood and wood materials* **1** 59
[6] Baraks A M, Nikifirov Yu N 1969 *Deep Treatment of Wood with the Incision Method* **1** 20
[7] Patyakin V I and Sokolova V A 2011 *Bulletin of KrasGAU* **5** 159
[8] Li X, Lu K, Lin L et al. 2010 *BioResources* **9** 13
[9] Dashti H, Shahverdi M, Taghiyari H, Salehpur S and Heshmati S 2012 *BioResources* **7** 3262
[10] Sheng He 2014 *BioResources* **9** 1482
[11] Liu H H, Wang Q W, Yang L, Jiang T, and Cai Y C 2005 *BioResources* **16** 3327
[12] Dömény J, Koiš V, Dejmal A 2014 *BioResources* **1** 372
[13] Daian G 2011 *Delamination in Timber Induced by Microwave Energy* **1** 142

Evaluating the mechanical properties using high beam energy electron energy loss spectroscopy

A Yu Igumenov and A V Dudarev

Reshetnev Siberian State University of Science and Technology, 31 Krasnoyarsky Rabochy Av., Krasnoyarsk, 660037, Russia

E-mail: igumenovau@mail.ru

Abstract. The electron energy loss spectra of Ge, Si, Cu, MgO, Al, Ni, C (Diamond) were analyzed. The possibility of mechanical parameters of various materials evaluating using bulk plasmon energy in electron energy loss spectra is shown. The bulk modulus of Ge, Si, Cu, MgO, Al, Ni, C (Diamond) is calculated. It is shown that bulk modulus logarithm has a linear correlation with the bulk plasmon energy.

1. Introduction

Determination of mechanical properties of nanostructures is an actual problem of modern materials science [1]. In the first instance, thin films, nanotubes and nanofibers are interesting for investigation [2]. Widely used techniques are acoustic microscopy [3], nanoindentation testing [4], microtensile testing [5], laser ultrasonics [6], micromechanical resonator-based sensing [7].

An original approach to the mechanical properties research of materials without its direct deformation was proposed in [8].This method is based on bulk modulus calculation using bulk plasmon energy. Bulk modulus allows calculating the Young's modulus, shear modulus, microhardness, Poisson's coefficient. This method was developed in [9-11], and applied to a wide class of materials.

In this work the electron energy loss spectra (EELS) of Ge, Si, Cu, MgO, Al, Ni, C (Diamond) were analyzed. The characteristic peaks energies were determined and its origin was interpreted. Bulk modulus of Ge, Si, Cu, MgO, Al, Ni, C (Diamond) were evaluated. The correlation analysis between the bulk modulus and bulk plasmon energy was carried out. It is shown that bulk modulus logarithm has a linear correlation with the bulk plasmon energy.

2. Experimental

According to [8], the bulk modulus B is approximately equal to the product of chemical bond type and squared bulk plasmon energy:

$$B = \alpha \cdot E_p^2, \tag{1}$$

where α depends on the type of chemical bonding, E_p – research material bulk plasmon energy. Two approaches can be used for the bulk modulus calculation:

$$B = 8.44 \cdot \left(\frac{m \cdot e}{h^2}\right)^2 \cdot E_p^2, \tag{2}$$

$$B = 7.29 \cdot N_{ve} \cdot \left(\frac{m \cdot e}{h^2}\right)^2 \cdot E_p^2, \quad (3)$$

where m – effective electron mass, e – electron charge, h – Planck constant, $N_{ve} = n \cdot V_{ws}$ – number of valence electrons, n – electron density, V_{ws} – the Wigner-Seitz sell volume. In this paper, formula (2) was used for calculation. The reference values of the bulk modulus B_{exp} are taken from [9].

EELS from open database [12], including results [13, 14], were analyzed. EELS of Si, Ge, Cu, and C were obtained with transmission electron microscope CM20 PHILIPS. Primary electron energy is 200 keV for Si, Ge, and 100 keV for Cu, C. EELS of Ni, MgO were obtained with transmission electron microscope CM30 PHILIPS at primary electron energy 300 keV.

3. Results and discussion

The initial integral EELS are presented in figure 1. The spectra obtained at high primary electron energies are chosen, since the surface excitations contribution to these spectra is minimal. It simplifies the analysis of bulk origin energy losses.

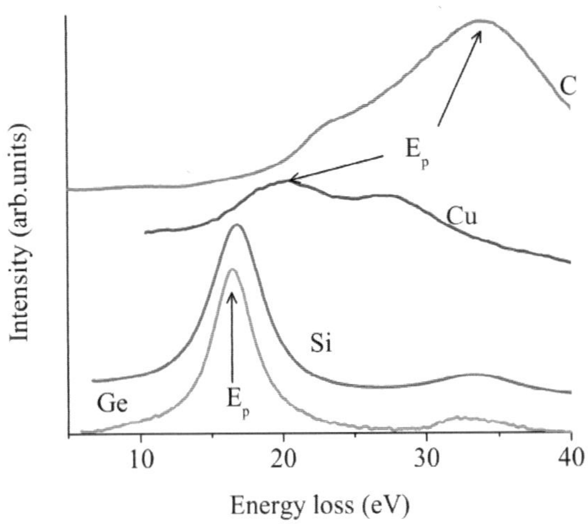

Figure 1. EELS.

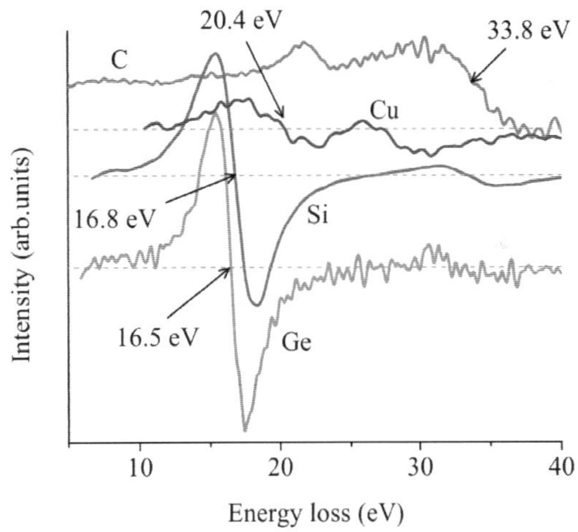

Figure 2. Differential EELS.

The energy loss spectra main maximum corresponds to the bulk plasma oscillations (E_p). In the spectrum of C, a low-energy region shoulder is observed. This shoulder is presumably related to a surface plasmon excitation [15]. The spectra (figure 2) were differentiated for more accurately bulk plasmon energy determination. The bulk plasmon energy and sturctureless background level are shown and marked in figure 2.

The obtained plasmon bulk energies are close to published data [9, 16] and shown in table 1. The table also shows the bulk modulus calculated with the bulk plasmon energy. It is shown that for most of the studied materials, a good agreement between the calculated and reference values of the bulk modulus is observed.

Table 1. Plasmon energy and bulk modulus.

Element	m	E_p (eV)	B (GPa)	B_{exp} (GPa)
Al	1.16	15.3	75.4	75.2
Ge	0.66	16.5	28.4	75
Si	1.12	16.75	84.31	97
Cu	1	20.4	99.69	137.8
MgO	1	22.0	115.94	155
Ni	1	28	187.81	177.3
C	2.12	33.75	1226.41	442

All calculated bulk modulus values are presented in figure 3 using logarithmic scale.

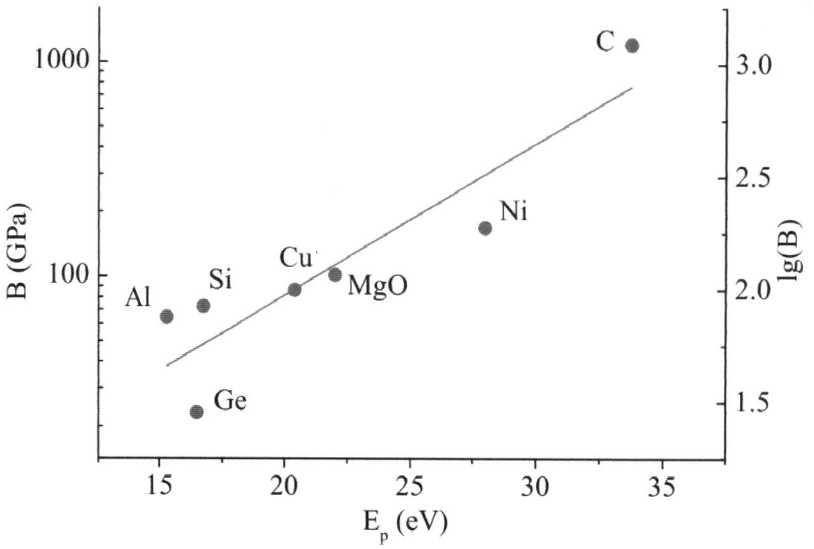

Figure 3. Bulk plasmon energy dependence of bulk modulus.

The dependence of the bulk modulus on the bulk plasmon energy is well approximated with the linear function $\log B = 0.07 \cdot E_p + 0.63$, and the correlation coefficient is $R^2 = 0.8$. Thus, it is possible to determine the elastic properties of materials using electron energy loss spectroscopy. This method can be promising in the development of elastic nanomaterials, the study of the mechanical properties of which is difficult because of its small size.

4. Conclusions

The electron energy loss spectra of Ge, Si, Cu, MgO, Al, Ni, C (Diamond) were analyzed. The characteristic peaks energies were determined and its origin was interpreted. The possibility of mechanical parameters of various materials evaluating using bulk plasmon energy in electron energy loss spectra is shown. The bulk modulus of Ge, Si, Cu, MgO, Al, Ni, C (Diamond) is calculated. It is shown that bulk modulus logarithm has a linear correlation with the bulk plasmon energy.

References

[1] Nysten B, Fretigny C and Cuenot S 2005 *Proc. SPIE Conf.* **5766** pp 78-88
[2] Kim Y Y 2017 *Materials.* **10(7)** pp 806-816
[3] Bamber M, Cooke K, Mann A andDerby B 2001 *Thin Solid Films.* **398** pp 299-305
[4] Pharr G M Curr. Opin. 2015 *Solid State Mater. Sci.* **19** pp 315-316
[5] Kihara Y, Nagoshi T, Chang T-F M, Hosoda H, Tatsuo S and Sone M 2015 *Mater. Lett.* **153** pp 36-39
[6] Zhang F, Krishnaswamy S, Fei D, Rebinsky D A and Feng B **2006** *Thin Solid Films.* **503** pp 250-258
[7] Mahmoud M A **2016** *Crit. Rev. Solid State Mater. Sci.* **41** pp 386–429
[8] Gilman J J 1999 *Philosophical Magazine Part B.* **79(4)** pp 643-654
[9] Oleshko V P, Murayama M and Howe J M 2002 *Microsc. Microanal.* **8** pp 350-364
[10] Oleshko V P and Howe J M 2005 *Mater. Res. Soc. Symp. Proc.* **839** pp P211.1-P211.6
[11] Oleshko V P 2012 *Journal of Nanoscience and Nanotechnology.* **12** pp 8580–8588
[12] Ewels P, Sikora T, Serin V, Ewels C and Lajaunie L 2016 *Microscopy and Microanalysis* **22(3)** pp 717-724
[13] Pantel R, Cheynet M C and Tichelaar F D *2006 Micron* **37** pp 657-665

[14] Ngantcha J P, Gerland M, Kihn Y and Riviere A 2005 *Eur. Phys. J. Appl. Phys.* **29** pp 83-89
[15] Raether H 1988 Surface Plasmons on Smooth and Rough Surfaces and on Gratings *Berlin, Springer-Verlag*
[16] Lifshits V G and Lunyakov Yu V 2004 Electron Energy Loss Spectra of Surface Phases on Silicon *Vladivostok: Dal'nauka*

IOP Conf. Series: Materials Science and Engineering 1230 (2022) 012007 doi:10.1088/1757-899X/1230/1/012007

Different approaches for SiO_2 inelastic electron scattering cross section spectra dissolving

A Yu Igumenov and T A Andryushchenko

Reshetnev Siberian State University of Science and Technology, 31 Krasnoyarskii rabochii prospekt, Krasnoyarsk, 660037, Russia

E-mail: igumenovau@mail.ru

Abstract. Inelastic electron scattering cross section spectra (Kλ-spectra) of SiO_2 were dissolved into bulk-loss and surface-loss components using factor analysis and Tougaard functions approximation. Primary electron energy dependences of various components intensities were received. It was shown that joint using of factor analysis and Tougaard functions approximation allows receiving more reliable results in inelastic electron scattering cross section spectra studies.

1. Introduction

Silicon dioxide is widely used for the development of micro- and nanoelectronics. SiO_2 is the most common gate dielectric for creating devices based on metal-dielectric-semiconductor (MOSFET) structures. Currently, various SiO_2-based materials are developing to achieve a higher band gap value. Accordingly, the elemental composition and chemical bonds analyzing methods, including electron spectroscopy methods, are in demand.

In this paper inelastic electron scattering cross section spectra ($K\lambda$-spectra) of SiO_2 were investigated by dissolving it into bulk-loss and surface-loss components by factor analysis (FA) [1-5] and Tougaard functions approximation [5-10].

2. Experimental

This paper presents the results of SiO_2 $K\lambda$-spectra investigation. Inelastic electron scattering cross section spectra ($K\lambda$-spectra) [11-13] are the product of the inelastic mean free path λ and the differential inelastic scattering cross section K (E_0, E_0 - E), where E_0 and E are the energies of primary and reflected electrons, $T = E_0 - E$ are electron energy losses. The spectra were analyzed at the primary electron energies of 300, 600, 1200, 1900, and 3000 eV.

Factor analysis allows processing the original set of spectra as a single array. Factor analysis identifies trends in spectra shape evolution with primary electron energy variation. It determines the number of independent components (Principal component analysis) from each other and divides them. As usually, these energy-loss components have bulk-loss and surface-loss origin.

Figure 1 shows the factor analysis stages. The original spectra is decomposed into the linear combination of most intense contributions. Components shape are optimized to reach Tougaard peak [11] shape using multiplying factor axes by the rotating matrix R (1) (Target transformation):

$$R = \begin{bmatrix} cos\varphi & sin\varphi \\ -sin\varphi & cos\varphi \end{bmatrix}, \quad (1)$$

where φ is the angle of factors rotation.

After that the original spectra are represented as a linear combination of factors.

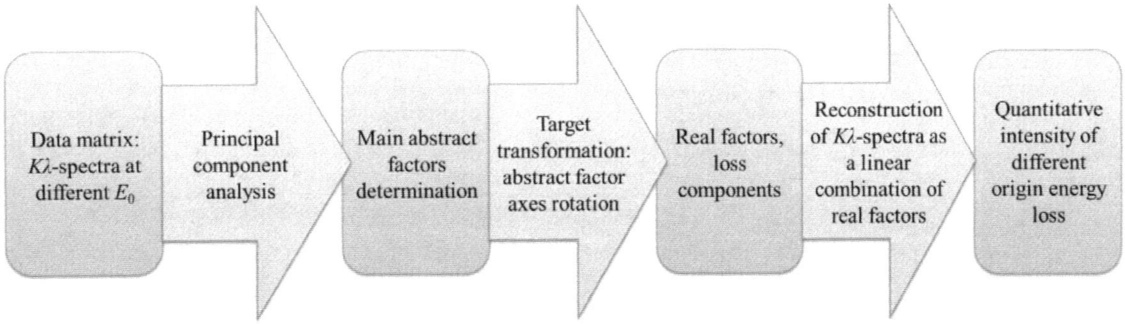

Figure 1. The factor analysis stages.

The Tougaard functions approximation is applied to each spectra separately and usually allows to bring out a larger number of components in spectra, then FA. It is possible to directly control the number of peaks, its width, position, and intensity. The three-parameter Tougaard function has the following form (2) [13]:

$$\lambda K = \frac{BT}{\left(C - T^2\right)^2 + DT^2}, \tag{2}$$

B, C, D are fitting parameters and have special values for different elements. The peak intensity depends on parameter *B*, peak energy depends on parameter *C* but does not correspond to it explicitly, peak width and, indirectly, peak energy depend on parameter *D*. Tougaard functions (universal classes of inelastic electron scattering cross section) are used to describe inelastic electron scattering cross section spectra.

The Tougaard three-parameter function has been modified (3):

$$\lambda K = \frac{BT}{\left(\left(T_p\left(-T_p + \sqrt{4T_p^2 + D}\right)\right) - T^2\right)^2 + DT^2}, \tag{3}$$

where the parameter T_p corresponds to the position of the peak maximum in absolute units.

3. Results and discussion

The original $K\lambda$-spectra were dissolved into 6 peaks (*a-f*) with using Tougaard function approximation. This number of components makes it possible to describe the original spectra with sufficient accuracy. It is possible to divide the contributions of different origin, which are bulk and surface plasmons, electron transitions combining certain peaks.

Using factor analysis two contributions were received. These contributions describe the different origin of electron excitations in a solid. The energy loss of the surface-loss factor which corresponds to the excitation of surface plasmon is 13.6 eV; the maximum of the bulk-loss factor is describing the excitation of bulk plasmon and located at energy loss 22.8 eV. The energy of the bulk-loss factor maximum is close to the bulk plasmon energy in SiO_2 [14].

The peaks and factors at primary electron energies 600 and 1900 eV are shown in figure 2.

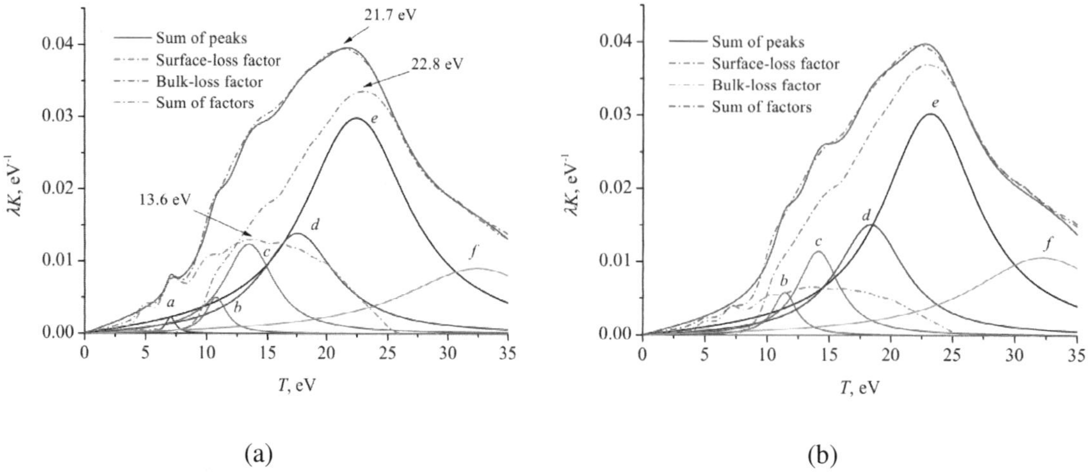

Figure 2. SiO_2 spectra dissolving at primary electron energies (a) 600 and (b) 1900 eV.

The Tougaard function approximation and factor analysis represent close to each other cumulative spectra. The fitting spectra shape is identical to the original $K\lambda$-spectra shape. This suggests that it is possible to accurately describe the original inelastic electron scattering cross section spectra and identify different origin contributions using both methods.

As it stands, the components intensity is changing with primary electron energy increasing. As can be seen in figure 2, the peak *a* disappears in the high primary electron energy spectra. Its position and behavior indicate that it belongs to the surface contribution group of the spectra. Figure 3 illustrates the primary electron energy dependences of peaks and factors positions.

Peak *a* is absent in the high primary electron energy spectra. The position of other peaks is almost unchanged that is an indicator of low mathematical uncertainty of these dissolving. The factor maximums energies remain constant when primary electron energy changes because the factors were obtained with using the entire array of source spectra, and its positions are determined from the general analysis. The position of the surface-loss factor is close to the position of Tougaard peak *c* from approximation set. The bulk-loss factor energy corresponds to peak *e* position. Figure 2 shows that these Tougaard peaks are the most intense. Thus, peaks *c* and *e* are interpreted as surface and bulk plasmons, and peaks *a*, *b*, *d*, and *f* are electron transitions.

The surface origin contribution intensity decreases and the bulk origin contribution intensity increases with an increasing of primary electron energy. This is well illustrated by the integral intensities (area) of factors or combinations of Tougaard peaks depending on the primary electron energy, as shown in figure 4. The combination of peaks *a*, *b*, and *c* describes the surface origin contribution in the inelastic electron scattering cross section spectra. Peaks *d*, *e*, and *f* belong to the bulk origin group of peaks.

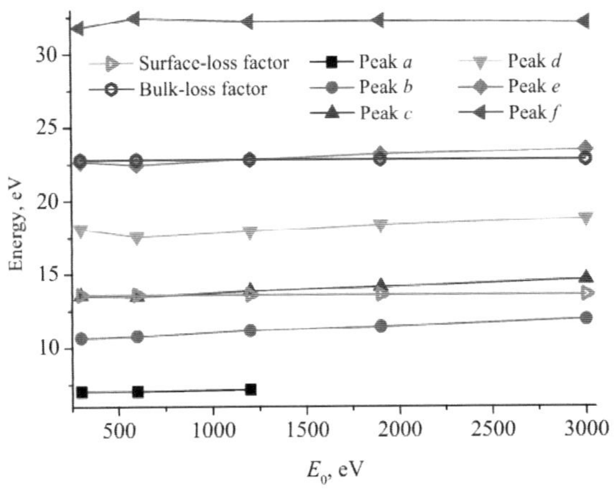

Figure 3. Primary electron energy dependences of peaks and factors positions.

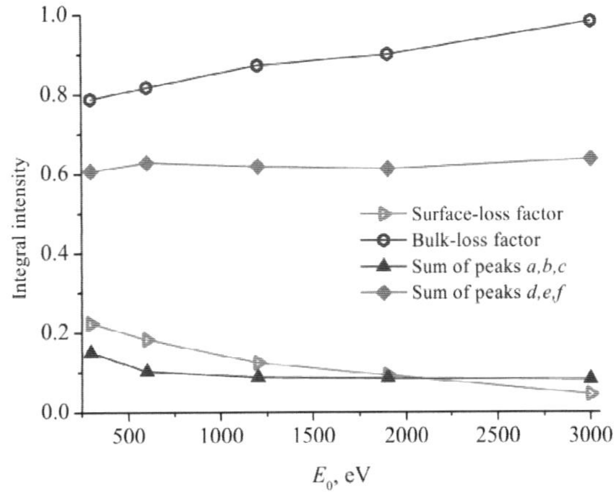

Figure 4. Primary electron energy dependences of peaks and factors area.

The behavior of peaks *a*, *b*, and *c* areas combination is similar to the dependence which is obtained for the surface-loss factor. This confirms the agreement of results that were obtained by different methods. Peaks *d*, *e* and *f* areas combination and the bulk-loss factor have an increasing character of the primary electron energy dependence that confirms its origin.

4. Conclusions

It is shown that the combined application of factor analysis and Tougaard function approximation allows obtaining more reliable and detailed information about the electron energy loss processes, determining peak energies, intensities and divide the bulk and surface contributions. The energies and integral intensities of surface and bulk plasmons and electron transitions are determined. The fitting peaks and factors energies are consistent with the electron energy loss literature data. Primary electron energy dependences of peaks and factors intensity indicate bulk and surface origin of excitations.

The authors express gratitude to Parshin A.S. for assignation the SiO_2 $K\lambda$-spectra.

References

[1] Jin H, Shinotsuka H, Yoshikawa H, Iwai H, Tanuma S and Tougaard S 2010 *Journal of applied physics.* **107** pp 083709 1-11

[2] Jin H, Yoshikawa H, Iwai H, Tanuma S and Tougaard S 2009 *Journal of Surface Analysis.* **15(3)** pp 321-324

[3] Igumenov A Yu, Parshin A S, Kanzychakova V O, Demin A M, Andryushchenko T A, Mikhlin Yu L, Pchelyakov O P and Zhigalov V S 2019 *IOP Conference Series: Materials Science and Engineering.* **467** pp 012019 1-7

[4] Igumenov A Yu, Parshin A S and Andryushchenko T A 2019 *Siberian Journal of Science and Technology.* **20(1)** pp 99-105

[5] Parshin A S, Igumenov A Yu, Mikhlin Yu L, Pchelyakov O P, Nikiforov A I and Timofeev V A 2015 *Semiconductors* **49 4** pp 423-427

[6] Parshin A S, Igumenov A Yu, Mikhlin Yu L, Pchelyakov O P and Zhigalov V S 2016 *Physics of the Solid State* **58** pp 908-914

[7] Parshin A S, Igumenov A Yu, Mikhlin Yu L, Pchelyakov O P and Zhigalov V S 2016 Technical Physic **61 (9)** pp 1418-1422

[8] Parshin A S, Igumenov A Yu, Mikhlin Yu L, Pchelyakov O P and Zhigalov V S 2016 *IOP Conference Series: Materials Science and Engineering.* **122** pp 012025 1-7

[9] Parshin A S, Igumenov A Yu, Mikhlin Yu L, Pchelyakov O P and Zhigalov V S 2017 *IOP Conference Series: Materials Science and Engineering.* **255** pp 012019 1-7

[10] Parshin A S, Igumenov A Yu, Mikhlin Yu L, Pchelyakov O P, Nikiforov A I and Timofeev V A 2014 *Vestn. Sib. Gos. Aerokosm. Univ.* **4 (56)** pp 230-235
[11] Tougaard S 1987 *Phys.Rev.* **35(13)** pp 6570-6577
[12] Tougaard S 1997 *Surface And Interface Analysis* **25** pp 137-154
[13] Tougaard S QUASES – Software packages to characterize surface nano-structures by analysis of electron spectra, available at: http://www.quases.com/
[14] Gritsenko V A, Petrenko I P, Svitasheva S N and Wong H 1998 *Applied Physics Letters* **72(4)** pp 462-464

Magnetoresistive effect in $Yb_xMn_{1-x}S$at small concentrations

A M Kharkov, H Abdelbaki, A V Gelgorn and N A Cheremnykh

Reshetnev Siberian State University of Science and Technology, 31 Krasnoyarskii rabochii prospekt, Krasnoyarsk, 660037, Russia

E-mail: khark.anton@mail.ru

Abstract. The results of measurements of electrical resistivity without a field and in magnetic field of 0.8 T in the temperature range 100 K < T < 450 K for composition$Yb_xMn_{1-x}S$ with x = 0.05, 0.1 are presented. For the x = 0.05 the gigantic magnetoresistive effect at room temperature is found.

1. Introduction

The relationship of magnetic and electrical properties is an important factor for creating electronic devices that operate on new principles and open new directions, such as spintronics [1-3] and new materials of multiferroics [4-9].

The interaction of electrons with an elastic lattice can also manifest itself in the resistive properties of semiconductors [10-13]. A strong magnetoresistive effect was detected in the vicinity of the orbital ordering in the two-orbit Hubbard model with ¼ the electron filling function at the site [14]. The resistance detects a small maximum in the region of formation of the orbital ordering [15], the electron density of states at the Fermi level splits in the magnetic field, which leads to an increase in the resistance in the paramagnetic phase [16-17]. The formation of an orbital ordering is accompanied by lattice deformations and changes in the magnetic state [18-20]. In chalcogenide compounds with polymorphic transitions, negative magnetoresistance is associated with tunneling of electrons having the same spin orientation in the magnetic field [21-23].

Manganese sulfides have a NaCl-type crystal FCC lattice with a constant unit cell parameter a = 0.5222 nm (MnS). Ytterbium sulfides have a NaCl-type crystal FCC lattice with a constant lattice a= 0.5693 nm (YbS). When the critical pressure value P = 8 GPa is reached, the YbS lattice is compressed by 12 % [24]. It can be expected that when replacing of manganese cations with ytterbium ions, the pressure of nearest neighbors will lead to a change in the valence of ytterbium ions and to the formation of a metal bond. If the valence of the metal to be +3 and the sulfur to be –2, then each unit cell containing four formula units of YbS will have four electrons that do not participate in the Me-S bond. These electrons will take part in the Me-Me bond and be collectivized. The formation of chemical bonds between ytterbium and manganese ions induces rearrangement of the electronic structure in a solid solution and changes in the magnetic and transport properties [25].

The purpose of this work is to determine the optimal conditions for the occurrence of the magnetoresistive effect by temperature and concentration. To achieve this goal, the temperature dependences of the resistance in the magnetic field will be measured and a correlation with the elastic system will be established from the temperature dependences of the coefficient of thermal expansion of the lattice.

2. Experimental results and discussion

X-ray diffraction analysis was performed on the DRON-3 installation. X-ray images and the crystal structure of $Yb_xMn_{1-x}S$ compositions were studied at room temperatures on the polished side surfaces of parallelepipeds: in the initial state after their preparation, after measurements of electrical measurement up to 500 K. The reflexes, different from the FCC structure are not observed. The size of a unit cell in a solid solution of $Yb_xMn_{1-x}S$ increases linearly with the concentration.

Electrical resistivity measurements were carried out using a four-probe compensation method at a direct current in the temperature range of 80 K – 500 K. Temperature dependences of electrical resistance for solid solutions $Yb_xMn_{1-x}S$ are shown in figure 1. The dependence of resistance on temperature has an activation form. They have a typical semiconductor type and do not differ qualitatively from the temperature dependence ρ (T) for MnS for compositions with $x \leq 0.1$. When heating a solid solution of $Yb_{0.05}Mn_{0.95}S$, the activation energy increases 1.7 times at T = 440 K. With increasing concentration the activation energy is decreased, and the temperature shifts to the low temperature range to T = 390 K for x = 0.1.

Replacement of manganese with ytterbium leads to an increase in the concentration of current carriers and a decrease in the activation energy. The absence of plateau in the temperature range 300 - 500 K of ρ (T) in the solid solution at magnetic field is observed. The energies of impurity states (E_i) are located below the bottom of the conduction band (E_c) by the value $\Delta E = E_c - E_i$ and with increasing concentration, the energy interval decreases.

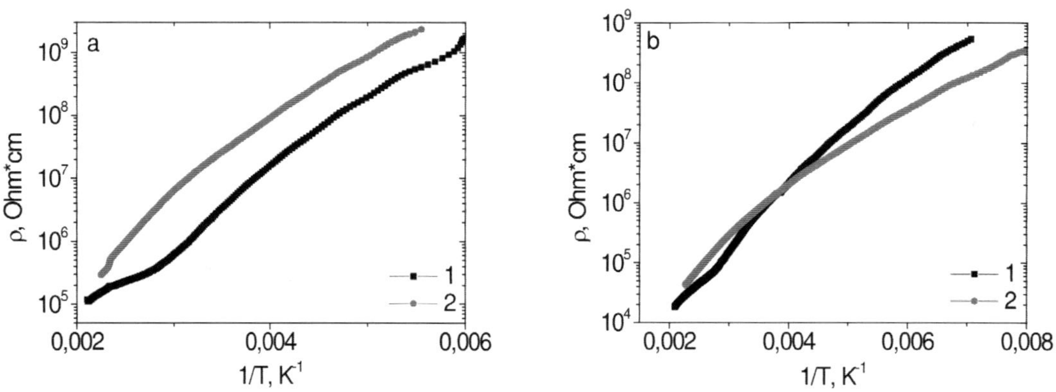

Figure 1. Temperature dependences of specific electrical resistivity for $Yb_xMn_{1-x}S$ samples with x = 0.05 (a), x = 0.1 (b) measured without a field (1) and in a magnetic field H = 0.8 T (2).

The influence of the magnetic field on the electrical resistance was studied without a field and in a magnetic field H = 0.8 T. In a solid solution of $Yb_xMn_{1-x}S$, the resistance increases in the magnetic field H = 0.8 T in the range of 150 K < T < 450 K and the relative change in the specific electrical resistance at magnetic field reaches a maximum at T = 329 K. In the area of room temperatures, have a giant positive magnetoresistive effect with a change in the resistance value by an order of magnitude (figure 2a). The value of the activation energy does not change in the range of 150 K – 300 K, and the preexponential is decreased tenfold in the magnetic field.

In $Yb_xMn_{1-x}S$ c x = 0.1, the activation energy of current carriers in the magnetic field decreases and the magnetoresistive coefficient changes sign from negative to positive at T = 260 K (figure 2b). The maximum value (ρ(H) – ρ(0)) / ρ(0) is observed at T = 360 K and the magnetoresistive effect disappears asymptotically at T = 500 K.

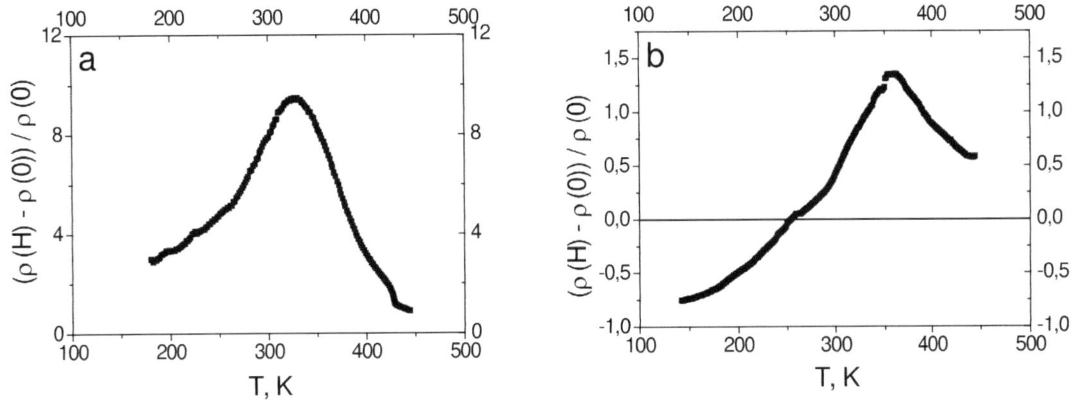

Figure 2. Relative change in electrical resistivity from temperature for samples $Yb_xMn_{1-x}S$ with x = 0.05 (a) and x = 0.1 (b).

The formation of an orbital ordering [26, 27] of the glass state type [28, 29] can lead to a change in elastic characteristics, both static and dynamic, for example, a change in the impedance in a magnetic field [30]. Below the coefficient of thermal expansion of the lattice as a function of temperature is studded.

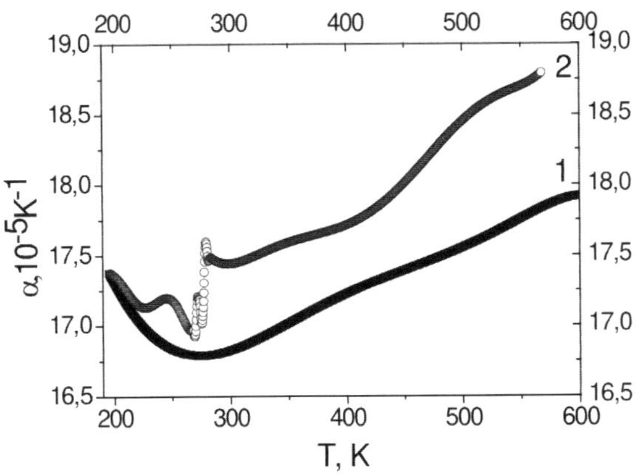

Figure 3. Temperature dependence of the coefficient of thermal expansion of the lattice for samples $Yb_xMn_{1-x}S$ with x = 0.05 (1) and x = 0.1 (2).

The thermal expansion coefficient was measured using a Netzsch dilatometer DIL-402C in the temperature range of 200 K – 750 K with a heating rate of 5 K/min. Fused quartz and corundum standards were used to calibrate and account for the thermal expansion of the measuring system. The results of strain studies (ΔL / L) and thermal expansion coefficient (α (T)) for samples $Yb_xMn_{1-x}S$ (x = 0.05, 0.1) are shown in figure 3. When heated, the coefficient of thermal expansion decreases and has a minimum at T = 275 K and a sharp jump in α (T) for the sample with x = 0.1. At this temperature, the permittivity and magnetic capacity have maximum values, and the magnetoresistive effect changes sign from negative to positive. Above T > 480 K, the slope in the temperature dependence α (T) (dα / dT = 0 at T = 480 K) decreases and the magnetoresistance disappears. The increase in the coefficient of thermal expansion with temperature may be caused by the anharmonicity of lattice vibrations as a result of electron-phonon interaction, which induces an asymmetry of the crystal field on the ion.

3. Conclusion

Thus, a magnetoresistive effect was found for $Yb_xMn_{1-x}S$ with x = 0.05 and x = 0.1 at temperatures above room temperature and for x = 0.05 the magnetic capacity sign changes in the vicinity of 200 K. A decrease in the activation energy of the transition of electrons from the impurity level to the zone of conductivity and mobility of charge carriers in a magnetic field was found. For a sample $Yb_{0.05}Mn_{0.95}S$, the magnetoresistive effect is 900 % at room temperature. The change of the magnetoresistance sign from negative to positive when heated in the sample region for $Yb_{0.1}Mn_{0.9}S$ is established. The critical temperature above which the magnetoresistance disappears is determined. The magnetoresistive effect is explained in the model of the orbital ordering of electrons. The change in the sign of magnetoresistance versus temperature is explained by the formation of magnetic and orbital ordering at different temperatures.

Acknowledgments
The reported study was funded by Grant of the President of the Russian Federation № MK-620.2021.1.2.

References

[1] Fert A 2008 *Rev. mod. phys.* **80(4)** 1517
[2] Kilian I and Jurgen S 2020 *Phys. Rev. B* **102** 054407
[3] Parish M M and Littlewood P B 2008 *Phys. Rev. Lett.* **101** 166602
[4] Hotta T and Dagotto E 2002 *Phys. B: Cond. Mat.* **312** 700
[5] Pyatakov A P and Zvezdin A K 2012 *Phys. Usp.* **55(6)** 557
[6] Volkov N V 2012 *Phys. Usp.* **55** 250
[7] Sherrington D 2013 *Phys. Rev. Lett.* **111** 227601
[8] Aplesnin S S, Udod L V, Sitnikov M N, Kretinin V V, Molokeev M S and Mironova-Ulmane N 2018 *Mat. Res. Exp.* **5** 115202
[9] Nohara Y, Yamasaki A, Kobayashi S and Fujiwara T 2006 *Phys. Rev. B* **74** 064417
[10] Aplesnin S S, Masyugin A N, Sitnikov M N and Ishibashi T 2019 *JETP Letters*, **110(3)** 223
[11] Mydosh J A 2015 *Rep. Prog. Phys.* **78** 052501
[12] Adepalli K K 2013 *Adv. Funct. Mater.* **23** 1798
[13] Khomskii D 2009 *Physics* **2** 20
[14] Peters R, Kawakami N and Pruschke T 2011 *Phys. Rev. B.* **83** 125110
[15] Farokhipoor S and Noheda B 2011 *Phys. Rev. Lett.* **107** 127601
[16] Aplesnin S S, Kharkov A M, Eremin E V, Romanova O B, Balaev D A, Sokolov V V and Pichugin A Yu 2011 *IEEE Trans. Magn.* **47(10)** 4413
[17] Romanova O B, Aplesnin S S, Udod L V, Sitnikov M N, Kretinin V V, Yanushkevich K I and Velikanov D A 2019 *J. Appl. Phys.* **125** 175706
[18] Moreo A, Yunoki S and Dagotto E 1999 *Science* **283** 2034
[19] Kugel K I and Khomskii D I 1982 *UFN* **136** 621
[20] Aplesnin S S 2003 *JETP* **97** 969
[21] Romanova O B, Aplesnin S S, Sitnikov M N, Kharkov A M, Masyugin A N and Yanushkevich K I 2019 *Solid State Comm.* **287** 72
[22] Miao B F, Huang S Y, Qu D and Chien C L 2013 *Phys. Rev. Lett.* **111** 066602
[23] Zeches R J, Rossell M D, Zhang J X, Hatt A J, He Q and Yang C H 2009 *Science* **326** 977
[24] Syassen K, Winzen H, Zimmer H G, Tups H and Leger J M 1985 *Phys. Rev. B* **32** 8246
[25] Aplesnin S S, Kharkov A M, Romanova O B, Sitnikov M N, Eremin E V, Gorev M V, Yanushkevich K I, Sokolov V V and Pichugin A Yu 2014 *JMMM* **352** 1-5
[26] Guo Y, Dai J, Zhao J, Wu C, Li D, Zhang L, Ning W, Tian M, Zeng X C and Xie Y 2014 *Phys. Rev. Lett.* **113** 157202
[27] Petrakovskii G A, Ryabinkina L I, Velikanov D A, Aplesnin S S, Abramova G M, Kiselev N I and Bobina A F 1999 *Phys. Solid State* **41(9)** 1520
[28] Choi Y J, Zhang C L, Lee N and Cheong S W 2010 *Phys. Rev. Lett.* **105** 097201
[29] Reichhardt C J and Bishop A R 2004 *Phys. Rev. Lett.* **92** 016801
[30] Aplesnin S S, Udod L V and Sitnikov M N 2018 *Ceramics International* **44** 1614

Determination of local heat transfer coefficient during boiling in annular channels of small diameters

A A Khodenkov, A V Delkov and E V Khodenkova

Reshetnev Siberian State University of Science and Technology, 31 Krasnoyarskii rabochii prospekt, Krasnoyarsk, 660037, Russia

E-mail: hodenkov.aa@gmail.com

Abstract. The presented work considers various methods for calculating the local heat transfer coefficient during boiling in annular channels of small diameters. The specificity of two different boiling modes, convective and nucleate, is analysed. The described calculation methods are compared.

1. Introduction

The main task in the calculation of convective heat transfer during boiling is to determine the heat transfer coefficient α. The main approach to determine heat transfer coefficients at present is the use of correlation dependences obtained experimentally. Due to the great scientific interest in studying the process of heat transfer during boiling, many experimental studies have been carried out. Unfortunately, numerous papers led to various confusions inadvertently. Currently, there are different correlations for calculating the heat transfer coefficient during boiling, but many of them have not been tested and verified in sufficient quantities. The aim of this study is to develop a reliable method for determining the local heat transfer coefficient during boiling in annular channels of small diameters.

The study of boiling processes is complicated by the presence of several boiling modes, the variety of structures of the two-phase flow during boiling, which determines the effect of the degree of dryness on the overall efficiency of the heat exchange circuit. It is necessary to note the problem of studying processes in the initial boiling regimes at low dryness degrees. In this case, the dominant mechanism of heat transfer intensification is the nucleate boiling mechanism, which leads to a sharp increase in the heat transfer coefficient. Such a change in the heat transfer coefficient has a significant effect on the heat transfer processes of two-phase cooling systems [1, 4].

For the boiling process with the directional movement of the coolant in pipes and channels, based on the research analysis of Russian and foreign authors, two main approaches based on different boiling models can be noted: convective boiling (CB) [5] and nucleate boiling (NB) [6].

2. Materials and research techniques

This study considers the applied technical problem of the heat transfer coefficient determining during boiling in annular channels of small diameters. This case is typical for evaporative heat exchangers that are part of various two-phase cooling systems. For a numerical assessment of the correlation dependences, we will take the following initial data as a test case: working fluid – R718 (water), pressure 47.622 kPa, boiling point 353,15 K, flow rate 0.010 kg / s, pipe with a diameter of 0.008 m.

As the theoretical basis of the study, an approach was chosen that defines the total heat transfer coefficient during boiling as the sum of two components: these are the components of the heat transfer coefficient for convective boiling α_{cb} and nucleate boiling α_{nb}. The summation is carried out taking into account the coefficients F and S:

$$\alpha = F \cdot \alpha_{cb} + S \cdot \alpha_{nb} \quad (1)$$

In turn, the parameters are determined by changing the Martinelli parameter. For the theory most common for boiling in a channel, Chen's theory, let us evaluate for the test case the dynamics of change and the interconnection of parameters: the Martinelli parameter; S parameter; F parameter; as well as the components of the heat transfer coefficient in convective boiling α_{cb}; and nucleate boiling α_{nb}.

The Martinelli parameter for given boundary conditions will be determined by the equation:

$$X_{tt} = \left(\frac{1-x}{x}\right)^{0.9} \left(\frac{\rho_v}{\rho_l}\right)^{0.5} \left(\frac{\mu_l}{\mu_v}\right)^{0.1} = 0.024\left(\frac{1-x}{x}\right)^{0.9} \quad (2)$$

where x – the dryness degree, ρ – the density, μ – the viscosity, v – the vapour phase, l – the liquid phase.

The F parameter for the given boundary conditions will be determined as:

$$F = 2.35\left(\frac{1}{X_{tt}} + 0.213\right)^{0.736} = 2.35\left(41.66 \cdot \left(\frac{x}{1-x}\right)^{0.9} + 0.213\right)^{0.736} \quad (3)$$

The F parameter characterizes the contribution of the heat transfer coefficient during convective boiling to the total value. This parameter for small values of x varies slightly, its most significant increase is observed at x> 0.5 (figure 1).

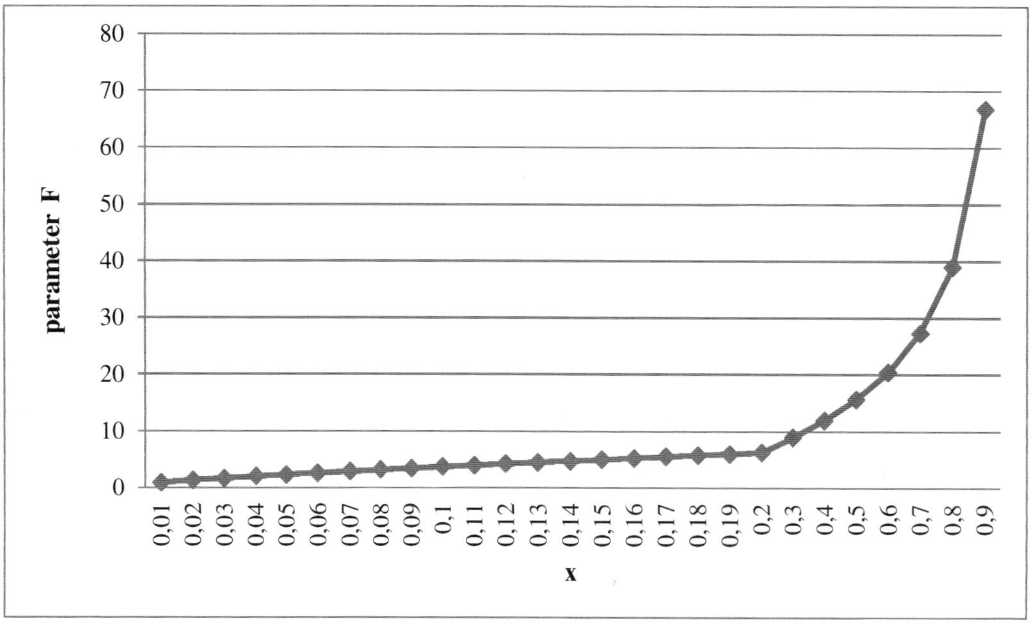

Figure 1. Change in F parameter depending on the dryness degree.

Therefore, the F parameter reflects the predominance of the convective boiling mechanism in areas at x> 0.5.

For given boundary conditions, we obtain the S parameter:

$$S = \left(\frac{1}{1 + 0.00000253\left[\mathrm{Re}_L F^{1.25} \right]^{1.17}} \right) \qquad \mathrm{Re}_\mathrm{L} = \frac{G(1-x)d}{\mu_L} \tag{4}$$

where Re_L – Reynolds coefficient for the liquid phase, G – mass flow rate, d – channel diameter, μ_L – liquid viscosity.

The S parameter reflects the nucleate boiling contribution to the total heat transfer coefficient graphically (figure 2): in accordance with its change, the mode of nucleate boiling prevails in the initial stages of boiling, which is then levelled.

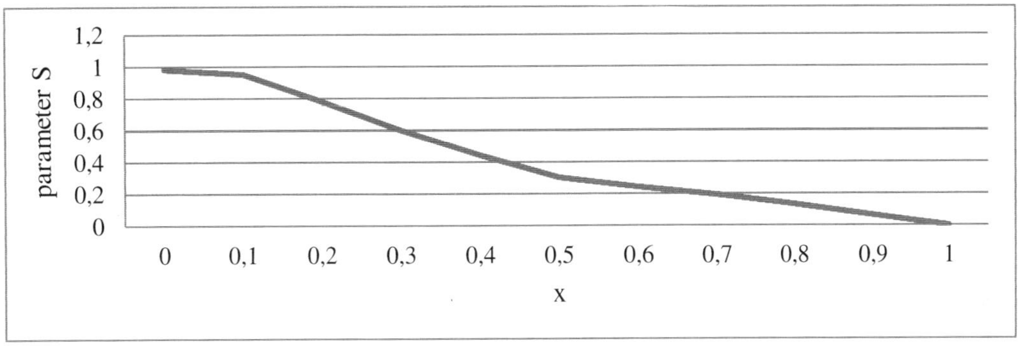

Figure 2. Change in S parameter depending on the dryness degree.

This coefficient has a maximum at the initial boiling points, and then decreases to zero at x> 0.8.
The heat transfer coefficient for convective boiling (F.W. Dittus) is determined by the equation [5].

$$\alpha_{cb} = 0.023 \mathrm{Re}^{0.8} \mathrm{Pr}^{0.4} \lambda / d \tag{5}$$

where Re – Reynolds criterion, Pr – Prandtl criterion, λ – liquid heat transfer coefficient, d – channel diameter.

Let us estimate the coefficient for given boundary conditions. The adjusted Reynolds criterion for the liquid phase according to the second formula:

$$\mathrm{Re} = \frac{Gd(1-x)}{\mu} \tag{6}$$

Taking into account $m = \rho F v$ $\quad \mu = \rho\eta$, $G = \frac{m}{F}$ (7)

$$\mathrm{Re} = \frac{\rho v d(1-x)}{\mu} = \frac{vd(1-x)}{\eta} \tag{8}$$

As can be seen from the above,

$$\alpha_{cb} = 0.023 \left(\frac{vd(1-x)}{\eta} \right)^{0.8} \frac{\lambda}{d} \mathrm{Pr}^{0.4} \tag{9}$$

The dependence of the heat transfer coefficient α_{cb} in the initial boiling sections is moderately decreasing depending on the dryness degree x character.

Further the heat transfer coefficient contribution at convective boiling to the total value of the heat transfer coefficient using the F parameter ($F \times \alpha_{cb}$) is estimated (figure 3). This parameter has a significantly increasing character, which is associated with an increase in the dryness degree x.

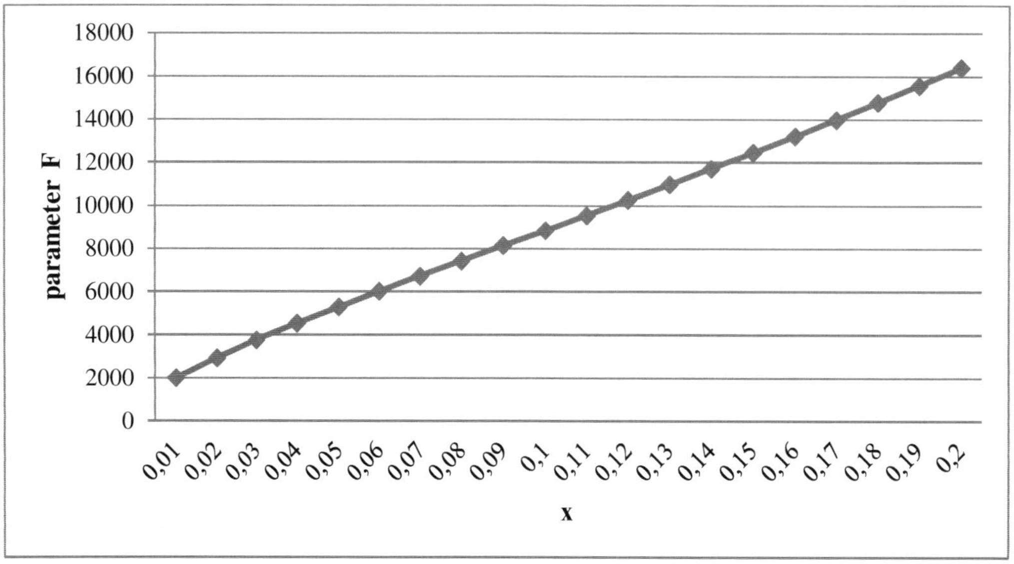

Figure 3. Convective component contribution to the heat transfer coefficient during boiling.

In general, the determination of the heat transfer coefficient for convective boiling from the dependences of J.C. Chen and F.W. Dittus does not cause significant difficulties and shows a graphic representation of the flow. This dependence can be used for the purposes of this study.

The possibility of using the M.G. Cooper dependence for nucleate boiling of heat transfer for the purposes of this study can be estimated as follows [3].

$$\alpha_{nb} = 55 \cdot \mathrm{P}_{\mathrm{red}}^{0.12} \cdot \left(-\log_{10} \mathrm{P}_{\mathrm{red}}\right)^{-0.55} \cdot M^{-0.5} \cdot q^{0.67} \tag{10}$$

where Bo - Boiling number, $\mathrm{P}_{\mathrm{red}}$ - reduced pressure.

Using the data of the test calculation task, it was obtained $\alpha_{nb} = 13600\left[W / m^2 K\right]$

It should be noted that when using the M.G. Cooper method, the definition of the S parameter will differ from the J.C. Chen method [2]. In this case, the S parameter is recalculated:

$$E = 1 + 24000 \cdot Bo^{1.16} + 1.37\left(\frac{1}{X_{tt}}\right)^{0.86} \tag{11}$$

$$S = \left[1 + 1.15 \cdot 10^{-6} \cdot E^2 \cdot \mathrm{Re}^{1.17}\right]^{-1} \tag{12}$$

A graph of the heat transfer coefficient distribution for the nucleate component of boiling by dryness degree x can be constructed taking into account the specified boundary conditions for the problem (figure 4).

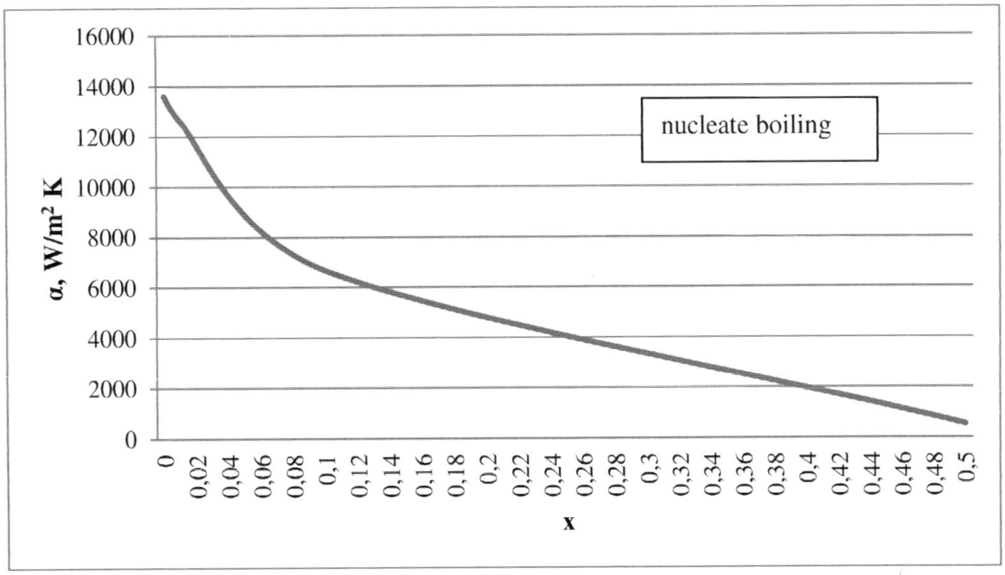

Figure 4. The effect of the nucleate component on the heat transfer coefficient during boiling.

3. Results and Discussion

For the test problem in this study, we determine the total contribution of the components from nucleate and convective boiling to the value of the heat transfer coefficient during boiling in the heat flow. The figure 5 presents the result of data comparison.

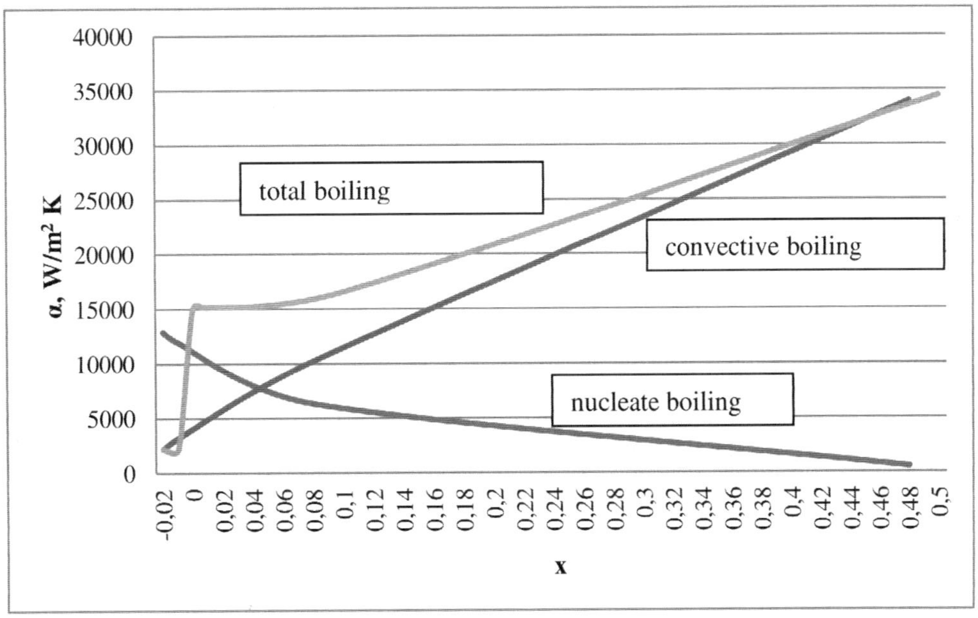

Figure 5. Total heat transfer coefficient depending on the dryness degree x.

In the range under consideration and according to the graph, it is observed: for nucleate boiling, at the beginning there is a sharp jump at the initial section (x = 0), then a proportional decrease down to zero; for convective boiling - proportional growth, starting with the base value of the heat transfer coefficient for the pure liquid flow. The overall pattern for the heat transfer coefficient distribution depending on the dryness degree x is as follows: at the beginning of the boiling section, there is a sharp increase due to the beginning of the nucleate boiling process. Then there is an insignificant reduction with a decrease in the heat transfer from nucleate boiling for dryness degree x = 0.05. After that, starting from x = 0.1, there is growth due to increase in convective boiling component.

To verify the obtained method a comparison of the calculated data with the experimental data in the test case was made. The figure 6 presents the result of this data comparison.

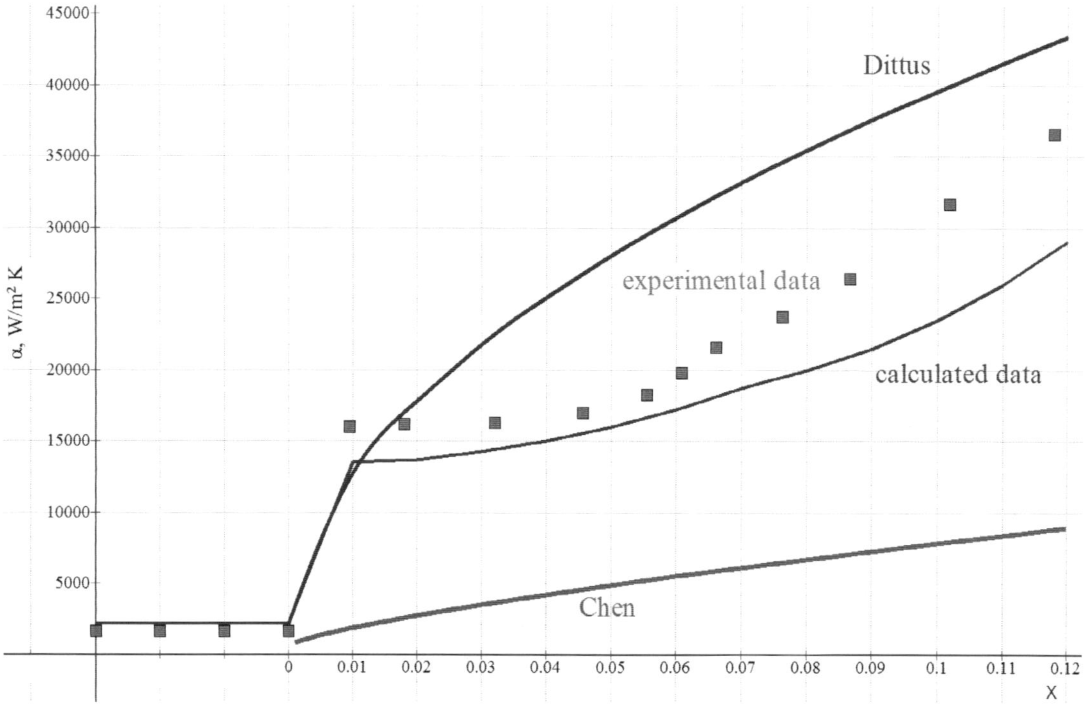

Figure 6. Calculated and experimental data comparison.

The resulting pattern is in good agreement with the experimental data obtained by the authors.

4. Conclusion

As a result of the present study, methods for determining the local heat transfer coefficients during boiling in annular channels of small diameters were obtained. The methodology is based on the convective and nucleate boiling modes components calculations. The main approaches to the calculation of these parameters were analysed and the most suitable methods for the problem under consideration were determined. The method was confirmed by experimental data for the test case.

References

[1] Cao E 2009 *Heat Transfer in Process Engineering* (McGraw-Hill Professional) p 576
[2] Chen J C 1966 *Correlation for Boiling Heat Transfer to Saturated fluids in Convective Flow* (I&EC Process Design and Development) pp 322–329
[3] Cooper M G 1984 *Saturated nucleate boiling. A simple correlation* (Proceedings of the 1st UK National Conference on Heat Transfer) **2** pp 785–793
[4] Dengler C E and Addoms J M 1956 *Heat Transfer Mechanism for Vaporization of Water Tube* (Chemical Engineering Progress Syposium Series) **52** pp 95–103

[5] Dittus F W and Boelter L M K 1930 *Heat transfer in automobile radiators of the tubular type* (Univ. Californ. Publ. Eng) pp 443-461
[6] Webb R L 1981 *The evolution of enhanced surface geometries for nucleate boiling* (Heat Transfer Eng. 2) pp 46–69

Investigation of the stress-strain state for the fuel tank separator

D A Klimovskiy and V Yu Zhuravlev

Reshetnev Siberian State University of Science and Technology, 31 Krasnoyarskii rabochii prospekt, Krasnoyarsk, 660037, Russia

E-mail: Klinsky92@yandex.ru

Abstract. The work carried out finite element modeling for the thin-walled shell of revolution loaded uniform pressure with using MSC NASTRAN software. The simulations results allow us to study the stress-strain state of the torus rolling zone for refine the boundaries of plastic deformation. It is determined that the region of plastic deformation does not arise in the entire torus zone. This region is limited by the outer part of the torus along the generatrix edges.

1. Introduction

To ensure proper supply fuel component from the tank to the liquid thruster with pressure feed system must be guaranteed separation of the liquid phase from the gaseous phase. Guaranteed separation is possible only by mechanical separation. Effective is the use of axisymmetric metal eversible separators [1].

Mathematical relationships, compiled on experimentally-based physical model of the plastic deformation for thin-walled shells of arbitrary form, is not accurate enough and do not correspond to the results of experimental data. This can be explained accepted without sufficient justification symmetrical borders for deformation plastic zone and the surface appearance is adopted as torus [2-4]. This approach requires for each form of the shell the procedure identification theoretical and experimental data for specifying coefficients of parameters eversion shell. In the shell, moving along the generatrix of plastic deformation during eversion, the thin-walled condition is not fulfilled and it is necessary to justify a new physical material model.

To reduce the cost of designing shells is necessary to clarify the physical model of plastic deformation for the thin-walled shell in the rolling zone. One of the methods for solving the problem to clarify the boundaries for deformation plastic zone of the torus for the separator can be computer simulation with using advanced finite element packages such as MSC NASTRAN.

2. Research model

Dimensions of the separator and a simulated three-dimensional image of the separator shown in figure 1. The loading scheme and the torus zone are presented in figure 2.

To conduct the study selected MSC NASTRAN computer simulation package. When working in similar software products, the calculation basis is the right application and registration of loads acting on the model, as well as creating high-quality finite element (FE) mesh.

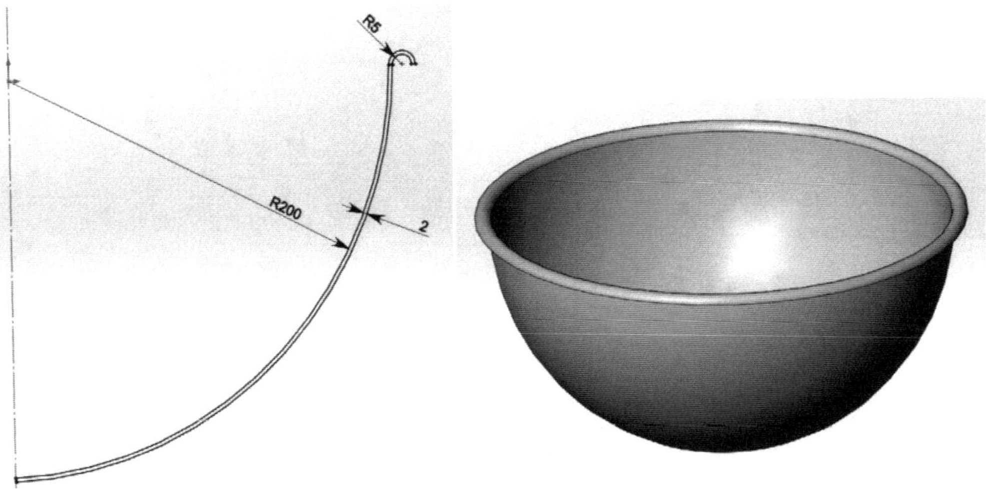

Figure 1. Dimensions of the investigated separator and 3D view.

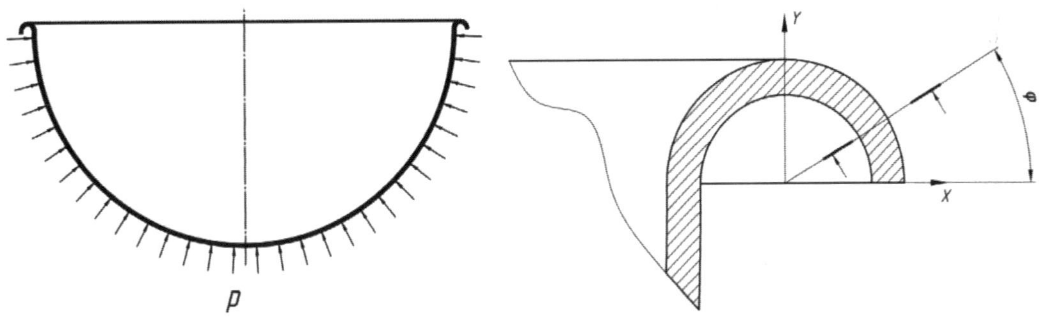

Figure 2. Separator loading scheme and torus zone.

Since posed problem is axisymmetric, it is advisable to use the Axisymmetric FE from section volume finite elements [5-7]. Axisymmetric FE is an annular member with a triangular or quadrangular cross-section. This type of FE is constructed in XZ plane from the base coordinate system, where Z - body rotation axis. For construction needs only half the cross section for the body of revolution. Model should not intersect the Z axis.

For high-quality meshing appropriateness simulate the cross-section of the separator finite elements without creating a geometric model.

The paper investigates two positions of the separator. Initial with a torus zone angle $\varphi=180°$ and intermediate with a torus zone angle $\varphi=120°$. Finite-element models are presented in figure 3.

Initial data for modeling:

Material - aluminum AD-1M ($E = 0.7 \times 10^5$, $\mu = 0.27$, $\sigma_{02} = 50$ MPa, $\sigma = 80$ MPa), isotropic, elastoplastic;

Load - $p = 0,1$ MPa , distributed on the bottom of the separator;

Boundary conditions - fixed edge of the outer shell;

IOP Conf. Series: Materials Science and Engineering 1230 (2022) 012010 doi:10.1088/1757-899X/1230/1/012010

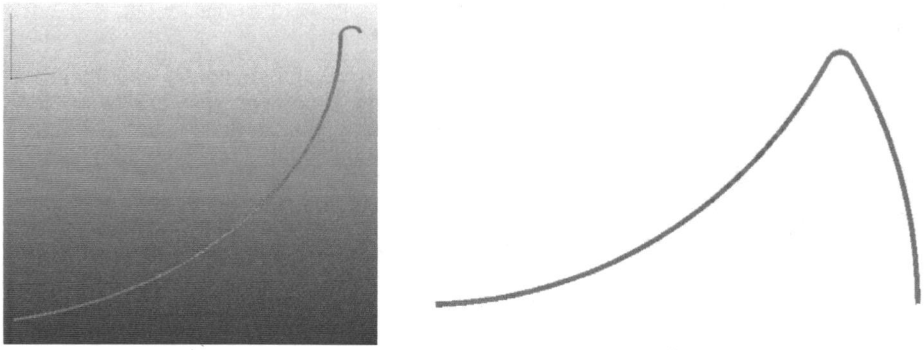

a) b)

Figure 3. Finite-element models of thin-walled shell: a) in the initial position; b) in intermediate position.

3. Results and Discussion

The simulation results for Von Mises stresses are presented in figure 4.

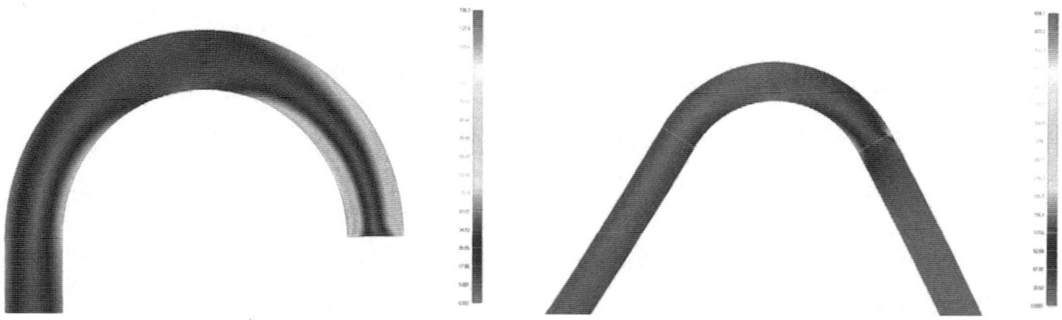

a) b)

Figure 4. Equivalent stress distribution: a) in the initial position; b) in intermediate position.

For greater clarity, we have limited the maximum displayed stress to 50 MPa (stress yield point of the material). We get the image in figure 5:

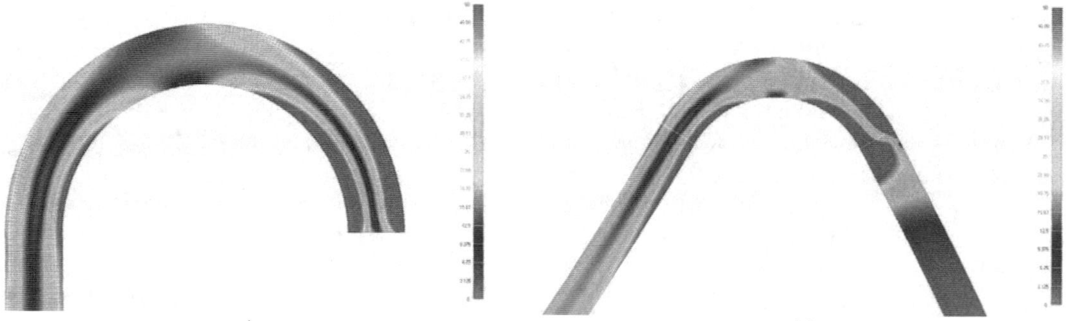

a) b)

Figure 5. Equivalent stress distribution with the restriction 50 MPa: a) in the initial position; b) in intermediate position.

The red zones in figure 5 displays where the stresses exceed the yield strength. Plastic deformations occur in these places.

The scaled deformed view is presented in figure 6.

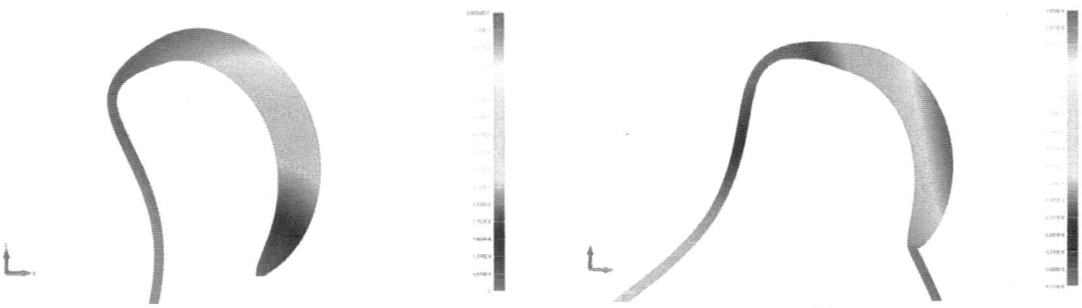

a) b)

Figure 6. The scaled deformed view: a) in the initial position; b) in intermediate position.

The analysis of stress-deformed state indicates that the plastic stress distributed in torus zone is not symmetrical relative to the minor axis of the torus and not uniformly. Plastic deformation is pronounced at the edges of the torus zone only in the outer layers of the shell (fig. 6). With the development of plastic deformations, the plasticity zones merge and the middle surface of the torus zone deforms at angles $\varphi = 0 \ldots 90°$.

The simulation results confirm that mathematical models of plastic deformation for thin-walled shells with an arbitrary shape of the generatrix, based on the symmetry of the boundaries deformation plastic zone, do not have sufficient accuracy.

4. Conclusion

1. A finite element model of two positions of the stress-strain state of the fuel tank separator in the MSC NASTRAN package based on axisymmetric finite elements has been created
2. Plastic stress distributed by torus zone is not symmetrical and uniformly.
3. Mathematical models of plastic deformation for thin-walled shells with arbitrary shape of the generatrix, based on the symmetry of boundaries deformation plastic zone, are not accurate enough.

References

[1] Gardymov G P, Parfenov B A and Pchelintsev A V 1997 *Technology of rocket science: a Training manual* (St. Petersburg: Special Literature) p 320
[2] Efremov V N, Zhuravlev V Yu and Yakubovich O P 2005 *Separators of fuel tanks with negative parallel deformation* monograph (SibSAU: Krasnoyarsk) p 76
[3] Kubrikov M V and Zhuravlev V Yu 2010 Design of diaphragm separators for elongated fuel tanks *Journal of SibSAU* (Krasnoyarsk) **4** 30 pp 105-107
[4] Zhuravlev V Yu, Klimovsky D A, Nazarova L P and Falkova E V Kinematics of thin-walled eversible shells during plastic deformation *Mechanics. Research and innovation.* (Gomel) **11** pp 75-79
[5] Rychkov S P 2013 *Structural modeling in FEMAP with NX NASTRAN* (Moscow: DMK Press) p 784
[6] Rudakova K N 2011 *FEMAP 10.2.0. Geometric and finite element modeling of structure* (Kiev: KPI) p 317
[7] Shimkovich D G 2003 *Structural Analysis in MSC NASTRAN for Windows* (Moscow: DMK Press) p 448

Study of the possibility of using biomass coniferous in the production of composite materials

A I Krivorotova and O A Usoltsev

Reshetnev Siberian State University of Science and Technology, 31 Krasnoyarskii rabochii prospekt, Krasnoyarsk, 660037, Russia

E-mail: tkmkai@mail.ru

Abstract. Processing of softwood biomass is a promising direction in the production of wood composite materials. Pre-treatment of waste debarking allows you to create chipboards with high thermal insulation and strength properties.

1. Introduction

Wood raw materials have a fairly complex and diverse structure and in cases where it is necessary to consider the mass of the entire tree, the term "biomass" is most often used.

In a live growing tree, there are three main parts: the crown, trunk, and root system. On average, the biomass of coniferous and deciduous species is distributed as follows: trunk-60-65 %, crown-15-20 %, roots and stump-15-30 % [1].

Standard logging technology separates the trunk of the tree from the crown and stumps with roots. All this remains in the woodlands. When using only stem wood, the loss of biomass in the logging process can be up to 35-40 %. In sawmilling, waste reaches 25-40%, and in woodworking-25 % or more.

In wood processing, the term "wood" refers to the bark-free tissue of the trunk, branches, and roots. At the same time, branches and roots are practically not used in the process of wood processing. In the total biomass, wood accounts for an average of 82 %, bark 15 % and tree greenery-3 %. For further use, the components of biomass need to be separated, since their structure and composition differ significantly from each other [2].

Coniferous species predominate in the forests of the Krasnoyarsk territory. The crown of coniferous trees is used very little. Needles are mainly used in the production of coniferous-vitamin flour. However, it should be noted that today whole directions of design application of needles are developing in the form of wall decorative noise-absorbing panels, biodegradable fabrics, furniture and carpets [3, 4]. Logging waste is practically not used, since its disposal is associated with a number of problems, including additional financial costs. Such waste is at best used as fuel. The main advantage of this fuel is its environmental safety, since its combustion produces a minimum amount of ash and harmful substances [5]. Along with pine needles, twigs and branches on the logging sites remains a large amount of waste wood cones. When felling a tree, the branches along with the cones located on them are cut off and remain in the cutting areas. In addition, with the increase in the volume of harvesting of coniferous seed stock in forest areas, the volume of waste generated after the extraction of seeds from cones increases.

Thus, the processing of softwood biomass is one of the most difficult problems in the complex of issues on the use of waste from wood processing, which is explained by the peculiar anatomical structure of needles, bark and cones. If we consider the tree bark, the relevance of this problem is only increasing every year. The bark is undoubtedly the raw material of the future and new technological processes will be developed on its basis, thanks to which the bark will be used productively and profitably. Solving the problem of industrial use of bark is important and is considered as one of the most urgent tasks [6].

Scientists from different countries are searching for optimal solutions for the use of bark as a raw material for the pharmaceutical, chemical industries, agriculture and the production of construction, decorative and slab materials.

The problem of bark recycling is also present in our region. Currently, the main method of recycling bark in large quantities in a relatively simple process is its combustion.

The use of bark in the production of slabs for various purposes has not yet become widespread. The lack of interest in the bark as raw material for producing slabs in the domestic industry is due to the presence of large resources of other higher quality wood waste (sawmill waste and wood, felling residues, wood wood, etc.), allowing to receive fibreboard and particle board with high physical-mechanical indicators.

Bark is considered as an additive to the main raw material in the production of particle board, fiberboard, cardboard. However, a lot of experience in the use of bark has been accumulated only in the production of particle boards. The lack of interest in bark as a raw material is mainly due to the fact that with existing production technologies, for example, wood fiber boards, the use of bark in the amount of 15% of the mass of absolutely dry fiber worsens the strength indicators of finished boards [7].

The results of processing the husk of cones, pine nut shells, and the remains of cones are also widely known. The most famous material is «Kedroplast». This is pressed into the resin of cedar-SAP-the remains of a cedar cone after extracting the seeds. It should be noted that cedar seed cones are used to produce a number of valuable products. There are quite a large number of works devoted to the complex processing of cedar nut biomass [8-12]. Processing of cones of other coniferous breeds is even less common and is conducted much less often. Most often, such processing is reduced to the use of cones as fertilizers (extracts, ash), mulch or fuel-smoking briquettes.

Despite the fact that numerous studies are currently being conducted and there are developments by foreign and domestic authors on the use of wood biomass, including the production of slab materials, biomass is processed poorly in industrial conditions. This fact makes it possible to develop new materials with special pre-defined physical and mechanical properties using various methods of preparing wood biomass.

2. Materials and research techniques

The purpose of the work was to determine the mode of pressing chipboard using a fine-grained mass of bark as a binder. Determination of the optimal mode of pressing chipboard based on bark was made according to the B^3 plan. Factors and their levels of variation are shown in table 1.

Table 1. Factors and levels of variation in the mode of pressing chipboard based on bark.

Name of the factor	Designation	The variation levels		
		upper	main	lower
		+1	0	-1
The pressure of pressing, MPa	P/X_1	1,4	1,0	0,6
Specific duration of pressing, min /mm	$\tau_{уд}/X_2$	2,5	1,5	0,5
Pressing temperature, ^{0}C	T/X_3	180	130	80

3. Results and Discussion

As a result of mathematical processing of experimental data, a regression equation was obtained that adequately describes the dependence of the strength limit for static bending on the compression mode of samples

$$\sigma = 0{,}773333 - 0{,}001 \times P + 0{,}081 \times \tau + 0{,}07 \times T - 0{,}0925 \times P \times \tau + 0{,}0325 \times P \times T + 0{,}0825 \times \tau \times T \quad (1)$$

Further processing of the experiment results allowed us to determine a promising pressing mode: P=1,4 MPa, τ = 2,5 min /mm, T = 180 °C.

Another important factor affecting the physical and mechanical properties of the developed plate is the ratio of components. The amount of finely ground bark varied from 20 % to 60 %. With an increase in the content of fine-ground bark mass in the chip mixture, there is an increase in the strength of manufactured samples, but at the same time, the difficulty of selecting the mode of pressing the chip composition increases. This is probably due to a significant increase in the vapor-gas pressure in the chip bag. When the humidity of the mixture is reduced by less than 20 % in order to reduce the vapor-gas pressure, the effect of "gluing" the chips is reduced. Samples are stratified and lose their shape stability. The static bending strength varies from 0,1 to 0,8 MPa when the material density changes from 220 to 400 kg/m^3. The plates were tested for water absorption by directly submerging the samples in water. With an increase in the amount of fine-grained bark in the material, the water absorption of plates decreases. Fine-grained bark particles have low hygroscopicity and wettability, almost repel water.

The result of the research can be considered the development of pressing modes and component composition for obtaining an eco-friendly chipboard based on finely fine-grained bark with physical and mechanical characteristics comparable to the characteristics of heat-insulating fibrolite or arbolite plates, sufficient for the use of this plate as a thermal insulation material.

4. Conclusion

As a result of the experiments, it was found that the density of the developed thermal insulation material, which provides quality indicators for thermal insulation boards, is from 220 to 400 kg/m^3. The recommended thermal insulation material has the following characteristics: thermal conductivity coefficient from 00,077 to 0,081 W/m·C°, water absorption in 24 hours – from 46 to 37 %. Optimal indicators of physical and mechanical properties have a material made on the basis of finely fine-grained bark and chips with a component ratio of 40/100.

References

[1] Terenteva E, Udovenko N and Pavlova E 2016 *Complex Chemical Processing Wood* (Saint Petersburg: Higher school of technology and energy) p 74
[2] Geles I and Koritskaya Z 1992 *The biomass of the tree and its use* (Petrozavodsk: Karel'skiy nauchnyy tsentr of the Russian Academy of Sciences) p 230
[3] Biodegradable furniture made of pine needles *Ecology*, available at: https://ecology.md/page/biorazlagaemaja-mebel-iz-hvojnyh-igolochek-foto
[4] How to make an acoustic panel from spruce needles on needles *Kuvalda.ru*, available at: https://www.kuvalda.ru/blog/articles/zolotye_ruky/kak-na-igolkah.html
[5] Runova E M and Ugryumov B I 1998 Complex processing of green conifers with the aim of obtaining biologically active substances *Cyberleninka*, available at: https://cyberleninka.ru/article/n/kompleksnaya-pererabotka-zeleni-hvoynyh-porod-s-tselyu-polucheniya-biologicheski-aktivnyh-veschestv
[6] Tsyvin M 1973 *The use of tree bark* (Moscow: Forest industry) p 24
[7] ZHuravleva L N and Devyatlovskaya A N 2007 Main directions of using wood waste *Cyberleninka*, available at: https://cyberleninka.ru/article/n/osnovnye-napravleniya-ispolzovaniya-drevesnyh-othodov/viewer
[8] Lebedeva O, Ushanova V, Rubchevskaya L, Repyakh S and Lobadina M 1995 Method of obtaining of cedar oil *Copyright certificate No* 2096443 RF

[9] Rubchevskaya L, Lebedeva O, Ushanova V, Devyatlovskaya A, Pronina L and Repyakh S 1996 Complex processing of pine nuts *Copyright certificate No 2138541* RF
[10] Lebedeva O, Ushanova V, Rubchevskaya L, Repyakh S 2000 Method of obtaining of cedar oil *Copyright certificate No* 2194745 RF
[11] Rubchevskaya L, Lebedeva O, Ushanova V, Liss E and Repyakh S 1999 Method for producing polyphenols *Copyright certificate No* 2174011 RF
[12] Khromov's Kedroplast *Green city*, available at: https://green-city.su/kedroplast-xromova

Processing red mud as a factor of removing the risk of environmental disasters: RUSAL experience

V I Litovchenko and N A Shumakova

Reshetnev Siberian State University of Science and Technology, 31 Krasnoyarskii rabochii prospekt, Krasnoyarsk, 660037, Russia

E-mail: litovchenkovi@yandex.ru

Abstract. The paper considers the current problem of waste management of aluminum production in the Krasnoyarsk Territory. The negative consequences of the existence of sludge collectors of aluminum production for the natural environment and the population of the region are shown. The analysis of the consequences of environmental disasters, such as cancer in the population, chemical pollution of groundwater and surface water caused by leakage of hazardous substances, is presented. The experience of RUSAL, as well as international experience in the field of waste processing is presented.

1. Introduction.

Currently, one of the pressing environmental problems in Russia is the problem of waste management. The composition of production waste becomes more componential, new substances appear that also have a negative impact on the environment, on human life and health. In order to reduce the amount of waste generated and reduce its negative impact on the environment, on human well-being and health, as well as in order to obtain secondary raw materials, energy or products with certain consumer properties, various methods of waste disposal are used.

The significant problem is the waste of aluminum production, which today is especially acute and, in fact, unpredictable in the Krasnoyarsk Territory, where there are aluminum giants that produce semi-finished products, both for their own needs and for further deeper processing. Among them are *RUSAL* Krasnoyarsk (primary aluminum) and *RUSAL* Achinsk (alumina production). A huge problem is the sludge accumulated over the entire period of their activity and already estimated in million tons. Their impact on the environment is enormous and is already comparable to a delayed environmental disaster. This is compounded by the fact that the disposal processes of these hazardous accumulations over decades of operation are only at the level under discussion. Due to the lack of efficient processing technologies, the bulk of the red mud is not used and is stored in special collectors. They have a negative impact on the environment [1].

The purpose of the paper is to draw attention to solving the acute environmental problem that arose in the Krasnoyarsk Territory in connection with the activities of aluminum production enterprises.

Placing storages of large-tonnage waste generated during the production of aluminum in large cities of the Siberian region (Krasnoyarsk, Bratsk, Irkutsk, etc.) poses a serious threat to the population and natural resources in the absence of effective methods for their disposal. In the aluminum industry, red mud (RM), the solid residue of bauxite and nepheline after hydrochemical processing and extraction of alumina, is the largest in terms of the amount of waste.

IOP Conf. Series: Materials Science and Engineering 1230 (2022) 012012 doi:10.1088/1757-899X/1230/1/012012

The world has accumulated more than 1.5 billion tons of red mud. Each sludge repository occupies 50-100 hectares of land 30–50 m high, contains tens of millions of tons of highly moist (60-90%) finely dispersed and plastic material [2].

Red mud is formed during the purification of bauxite (the main raw material for the production of aluminum) in the production of alumina in the so-called Bayer process (the process for producing pure aluminum oxide). The red mud, from which aluminum oxide is removed, is contaminated with alkali and therefore poses a danger to the environment and humans: when it comes into contact with the skin, it begins to corrode. For each ton of aluminum oxide produced, 360 to 800 kg of sludge is produced [3].

The existence of the OAED facilities (objects of accumulated environmental damage), which are also the sludge collectors of aluminum production, is fraught with serious negative consequences for the natural environment and the population. Adjoining lands and surface water bodies, as well as waters under objects, are polluted by a wide variety of organic and inorganic substances, dangerous microorganisms, the set of which determines the type of activity that generated these objects. Among the pollutants of organic and inorganic nature are for example, heavy metals, petroleum hydrocarbons of various degrees of transformation, as well as poorly soluble chemical compounds in water, among which the most dangerous are polycyclic aromatic hydrocarbons, polychlorinated aromatic compounds [4].

2. Materials

Currently, Russian aluminum smelters generate a significant amount of solid waste that is hydraulically directed to sludge fields. The latter are hydraulic structures, inside the dams, the wastes are washed, and the liquid part of the pulp is separated and returned to the sludge disposal system. It is known that the consequences of the placement and storage of waste in sludge fields and landfills lead to high costs in the operation of sludge fields, cause the loss of valuable components for electrolysis (F, Al, etc.), and also require compensation for significant environmental damage. Environmental damage is caused, first of all, by chemical pollution of groundwater, which occurs as a result of filtration losses of industrial water, introduced into natural aquifers. During the operation of aluminum smelters, each of them has accumulated gigantic volumes of finely dispersed fluorine, sulfate and carbon-containing wastes, which are a mechanical mixture of finely dispersed dust from electrostatic precipitators, gas treatment sludge and flotation tailings [5].

Enormous environmental damage is also caused by disasters such as in Hungary near the town of Ajka, where one of the largest man-made disasters occurred in the 21st century and in the history of the country. As a result of the destruction of the toxic waste tank at an aluminum plant 160 km west of Budapest, about 700 thousand cubic meters of red mud were spilled [6].

The sludge, which is waste produced during aluminum manufacture, swept cars off roads and damaged bridges and homes, forcing the evacuation of 400 residents. About 7,000 people have been directly affected by the spill.

So far, about 1m^3 of sludge has leaked from the reservoir. Seven towns, including Kolontal, Devecser and Somlovasarhely, have been affected.

Doctors said that the injured were being closely monitored because the chemical burns caused by the sludge could take days to emerge and what could seem like superficial injuries could later cause damage to deeper tissue [7].

The government declared a state of emergency in three areas.

Hundreds of millions of tons of toxic aluminum waste, similar to those that fell into the Hungarian rivers and literally poisoned the lives of tens of thousands of people, have been accumulated in Russia to date. This waste, the so-called red mud, according to environmentalists, makes the aluminum industry the most dangerous industry in the world [8].

The risks of a recurrence of such a catastrophe increase due to increased natural disasters: earthquakes, torrential rains and floods, as well as terrorist attacks. But they can be completely eliminated by replacing the storage of red mud with its full processing. However, the developed processing technologies are not implemented due to the lack of guarantees of the return on investment in the construction of shipping complexes. Until the processing problem is resolved, disasters will continue.

On January 31, 2013, an emergency occurred at the Achinsk Alumina Refinery. Due to the breakthrough of the sludge "lake" of the Achinsk Alumina Refinery (AAR) - the RUSAL plant and the increase in the drainage water level in the bypass channel of the AAR sludge maps, a short-term overflow of water onto the adjacent territory occurred. The incident occurred on the evening of January 31, but the information about it became known only a few days later. According to the information of the Krasnoyarsk news agency Press-Line, the management of the plant until the last did not allow representatives of the Ministry of Emergencies or employees of Russia's Ros Prirod Nadzor environmental watchdog to the accident site. At the same time, according to the publication, the breakdown of the sludge pipe and overflow of the bypass channel of the AAR led to flooding of the area around the sludge fields of the plant with alkaline filtrate. Emergencies at the AAR happen regularly. A year earlier there was an explosion, even earlier - the fish had a rest in the nearby body of water, now - the leak of alkaline solution from the sludge fields. The last incident was dangerous in that the alkali destroys all life in the territory where it has already entered, and there was also a threat of getting into the Chulym River [9].

Speaking about the Krasnoyarsk Territory, scientists at the Siberian Federal University recall that the most catastrophic situation with waste from the metallurgical industry has developed in Achinsk. Over the decades, a huge slurry mountain has formed there. The main danger of these sludges is that they are highly alkaline. The ph factor is 10-13 units. When it enters water, a process such as an increase in the alkaline factor of water occurs. This can cause burns, death of plants, microorganisms and even animals. As a result of the spill of slurry fields, the lands become lifeless, resembling a Martian landscape. Achinsk spends a lot of money on the reconstruction of sludge fields, maintaining their performance, but leaks occur periodically. The last leak was in 2015. The accident is not major, but the consequences are quite serious. And according to the results of the investigation, the environmental prosecutor's office fined Achinsk alumina refinery for 408 million rubles. The problem is that a huge amount of waste is generated in alumina refining, much more than in aluminum production. According to RUSAL, this is about 12 million tons of waste. It is processed no more than 10%. Most of the waste is stored in sludge fields. Stimulation of the processing of sludges begs, because around Achinsk there is already not enough agricultural land, because many of them are occupied by landfills, by sludge cards.

The second problem is the sludge fields of the Krasnoyarsk aluminum smelter, the sludge of which is harmful because of fluorine. Fluorine intoxication leads to cancer, thyroid disease, and a number of other serious diseases. Despite the fact that special measures are being taken to retain sludge, studies by the Siberian Federal University scientists have shown that the fluorine content is very high in snow. The studies were conducted in 2015. Fluorine contaminated most of the Sovetsky district of Krasnoyarsk, part of the Central district of Krasnoyarsk, and the right bank of the Yenisei River. The fluorine content is much higher than permissible.

The conclusion suggests itself that fluorine from sludge fields is spread with the help of the wind, which leaks and enters the atmosphere in the form of a dust fraction. Unfortunately, this issue is poorly researched. Initial studies show that the situation is quite dangerous. The 4-5 times excess was fixed at such districts of Krasnoyarsk as Severny and Vzlyotka. This is quite a serious and a dangerous factor. An increase in the incidence of cancer occurs every year, and oncology appears not only in the older generation, but also among young people and children. Then there are diseases of the endocrine system and blood circulation. Statistics show that residents of areas adjacent to the plant are often sick.

3. Results and discussion

Despite the rather long-term operation of aluminum production in the Krasnoyarsk Territory (RUSAL Achinsk and RUSAL Krasnoyarsk), the issue of disposal of accumulated hazardous waste in sludge storage facilities has not yet been resolved. At AAR, about 100 million tons of nepheline sludge is currently stored on sludge collectors, with an annual increase of 4 million tons. At the Krasnoyarsk aluminum smelter, another sludge program No. 5 has been launched, which is ready to accommodate about a million tons of hazardous waste. These waste disposal sites carry a dangerous technological pressure on the environment, since they are located in the urban areas of the cities of Krasnoyarsk and Achinsk. It is necessary to introduce technologies for the deep processing of these wastes. In fact sludge fields should be completely removed from Krasnoyarsk.

Krasnoyarsk chemists, technologists have already found how to recycle and where to use recycled waste from the Krasnoyarsk aluminum smelter. Research and calculations were carried out on their own initiative, without the support of RUSAL. Doctor of Chemistry, professor of Siberian Federal University Boris Kulikov said that from the sludge it is possible to obtain additives necessary for the production of cement - suitable in construction and absolutely safe for human health. The construction of a processing plant requires 15-20 million dollars. This will allow Krasnoyarsk aluminum smelter to enter non-waste production [10].

RUSAL Krasnoyarsk Company pays great attention to reducing the negative impact of the operation of slurry storage facilities on the environment. In recent years, reconstruction of hydraulic structures and screening of sludge maps has been carried out, as well as the reconstruction of a system for the removal of sludge water, which allows withstanding sludge moisture in the range that reduces the filtration of contaminated industrial water into groundwater. Simultaneously with the reconstruction, monitoring of the status of groundwater is carried out. The existing operational network of ecological monitoring wells allows to control the impact of waste disposal sites on groundwater.

The technical director of RUSAL, Viktor Mann, said that the creation of a technology for the recycling of waste was an important area of work. So, at the Krasnoyarsk aluminum smelter (KrAZ), a two-stage gas purification section was commissioned to produce sodium sulfate in the first phase. It is used in the production of cellulose, glass, detergents, as well as building materials. Investments in this project amounted to $ 1.14 million. The new technology developed by the RUSAL Engineering and Technology Center surpasses foreign analogues in a number of ways: there is no discharge of liquid and solid wastes. In addition, this installation, which removes the substance from the gas treatment, reduces the maintaining equipment cost. "The two-stage gas purification with the removal of sulfur in the form of a commercial product at the Krasnoyarsk aluminum smelter has no world analogues. The effectiveness of the technology is above 99.5% for fluorine and 96.7% for sulfur, while world analogues do not exceed 92%. We have already launched the same project at the Novokuznetsk aluminum plant. The plans are to start using this technology at the Irkutsk and Bratsk factories"said Victor Mann. The introduction of the new technology will allow RUSAL to abandon the construction of new slurry fields at KrAZ, which will save up to $ 1.39 million per year. The plant management understands that the methods for handling sludge - slag-like waste from electrolysis production - are one of the most pressing environmental problems in the production of aluminum [11].

4. Conclusion

Taking into consideration above information, processing red mud could remove the risks of major environmental disasters in and around alumina refineries. In recent years, a number of proposals have been made and technical developments have been completed on the technologies for utilizing both sludge accumulations and sludge coming from the process stream. In laboratory conditions, technologies for pelletizing fine sludge with transfer to the iron and steel industry and cement production were developed and tested on small pilot plants. They gave positive results, but, unfortunately, RUSAL did not rush to invest in large-scale production of sludge utilization. It is easier and cheaper to build new maps, to reclaim waste sludge collectors, and to pay fairly small environmental fees for negative environmental impacts. There is practically no incentive for recycling, but not the accumulation of hazardous waste from the authorities. Agricultural land is contaminated, hazardous substances get into groundwater. Unfortunately, this dangerous impact on the environment and human health is currently underestimated. This is an environmental time bomb, set aside for future generations.

References

[1] Litovchnko V I, Misuna I I and Shumakova N A 2020 Contemporary analysis of Krasnoyarsk environmental problems *IOP Conference Series: Materials Science and Engineering* **822**

[2] Trushko V L, Utkov V A and Bazhin V Yu 2017 The relevance and possibilities of the complete processing of red mud from alumina production *Records of the Mining Institute* **227** pp 547-553

[3] Boyarintsev A, Aung M and Aung K, Stepanov S 2018 Aluminium leaching from red muds for their complexing reprocessing *Proceedings of VSUET* **3** pp 317-322

[4] Solovyanov A A and Chernin S Ya 2017 *Elimination of accumulated environmental damage in the Russian Federation* (M.: Science RAS) 456 p

[5] Ozersky G F and Botvich G F 2008 The influence of solid waste from primary aluminum production of JSC Rusal Krasnoyarsk on the quality of groundwater *Bulletin of the Peoples' Friendship University of Russia. Series: Engineering Research* **3** pp 98-102

[6] Hungary covered with red mud *Kommersant* **185**, available at: https://www.kommersant.ru/doc/1516625

[7] Hungary toxic sludge spill an «ecological catastrophe» says government *The Guardian*, available at: https://www.theguardian.com/world/2010/oct/05/hungary-toxic-sludge-spill

[8] Volkova O N 2016 Environmental problems of waste disposal at the White Sea sludge collector of the Kaprolactam plant of Sibur-Neftekhim *Bulletin of the University of Minin* **1**

[9] The authorities of Achinsk assure the townspeople that the breakthrough of the sludge collector of the Rusal smelter is not dangerous *Polit.ru.*, available at: https://polit.ru/news/2013/02/05/ekologia_achinsk/

[10] SFU scientists sound the alarm: Ecological bomb will explode due to KrAZ waste in Krasnoyarsk *Bomba News*, available at: https://bomba.news/news/7685/

[11] How metallurgists get rid of the «environmental backpack» *Nezavisimaya Gazeta*, available at: http://www.aluminas.ru/media/media-publications/aluminium_in_its_pure_form/

IOP Conf. Series: Materials Science and Engineering 1230 (2022) 012013 doi:10.1088/1757-899X/1230/1/012013

Particularities of the interstitial atoms and vacancies clusters formation in a thin cadmium telluride foil during in situ electron irradiation in a TEM

Y Y Loginov[1,2], A V Mozzherin[2], N N Paklin[2]

[1]Reshetnev Siberian State University of Science and Technology, 31 Krasnoyarskii rabochii prospekt, Krasnoyarsk, 660037, Russia
[2]Siberian Federal University, 79 Svobodny Av., 660041 Krasnoyarsk, Russia

E-mail: loginov@sibsau.ru; Amozzherin@sfu-kras.ru

Abstract. The formation of interstitial atoms and vacancies, as well as their clusters in the form of dislocation loops and voids in CdTe is simulated. The sizes and features of the growth of dislocation loops and voids were determined depending on the irradiation time, taking into account the decrease in the number of nodes of the semiconductor crystal lattice with the irradiation time, since in experiments we studied a thin CdTe foil in a transmission electron microscope (TEM). The calculated and experimental data are compared.

1. Introduction

Cadmium telluride (CdTe) relating to compounds A_2B_6 is one of the main materials of semiconductor microelectronics. This material is promising for the production of gamma and x-ray detector devices, new hard x-ray detection systems for space science, small pixel imaging detectors, ultra-high-resolution SPECT with CdTe detectors [1-5]. However, upon irradiation of CdTe, in particular by electrons, the structural defects are formed in it [6, 7], which affect the electrophysical properties of this material and devices made on its basis.

Computer simulation of the interstitial atoms and vacancies clusters formation in cadmium telluride upon irradiation with electrons was carried out in [7, 8]. In real experiments on the analysis of defect formation upon electron irradiation in a TEM, a thin foil is studied and the formation of clusters of point defects in this foil over time tends to decompose the material [9]. This means that in a thin foil, when modeling the processes of defect formation in TEM, it is also necessary to take into account the change in the number of crystal lattice sites of the material under study, which decreases with the duration of irradiation.

2. Results and Discussion

In previous studies, it was found that the types of defects, their concentration and the rate of defect formation in semiconductors depend on the stacking fault energy (SFE) [6, 10]. One of the options for the formation of dislocation loops is the presence of partial dislocation loops. TEM studies have shown that Frank and Shockley dislocations are involved in these processes [10]. The results made it possible to establish the “critical” radii of dislocation loops r_{crit}, which determine the boundaries that separate processes with different qualitative and quantitative changes in the evolution of a defective network of cadmium telluride. It was assumed that the value of r_{crit} is a measure of the material's resistance to defects.

To determine this parameter, the classical model describing the energy of the formation of Frank loops was upgraded, taking into account that the formation of this loop occurs from partial Shockley dislocations, which is more energy-efficient.

$$E_{(l,perfect)} = E_{(SF)} + E_{(l,Frank)} + 2E_{(Shockley)},$$

$$E_{SF} = \pi r^2 \gamma,$$

$$E_{l,Frank} = 2\pi r \frac{\mu b_{Frank}^2}{4\pi(1-\upsilon)}\left[\ln\left(\frac{8\alpha r}{b_{Frank}}\right)-1\right],$$

$$E_{Shockley} = 2\pi r \frac{\mu b_{Shockley}^2}{4\pi}\left[\cos(\beta)^2 - \frac{\sin(\beta)^2}{1-\upsilon}\right]\ln\left(\frac{\alpha r}{b_{Shockley}}\right).$$

As a result, we get:

$$\gamma = \frac{\mu a^2}{4800\pi r_{crit}(1-\upsilon)}\left[400\upsilon\ln\left(\frac{\alpha r_{crit}}{a}\right)+32+343\upsilon\right],$$

where $E_{l,\,perfect}$ is the total energy of dislocation loop formation, E_{SF} is the energy of the stacking fault, $E_{l,\,Frank}$ is the elastic energy of the dislocation line, $E_{Shockley}$ is the energy of the partial Shockley dislocation, μ is the shear modulus, υ is the Poisson coefficient, α is the dislocation core factor, a is the lattice parameter, γ is the total energy of the stacking fault, b is the module of the Burgers vector, r is the radius of the dislocation loop.

The values of some parameters and r_{crit} for CdTe "generalized model" are shown in table 1.

Table 1. The values of some parameters and r_{crit} for CdTe "generalized model".

Sample	lattice parameter, *nm*	Poisson coefficient [11]	Shear modulus, *Pa* [11]	SFE, J/m^2 [6,10]	Critical radius, *nm*	
					Classic model	Classical model taking into account Shockley dislocations
CdTe	0.648	0.41	$9.2\cdot10^9$	$(11 \pm 1.9)\cdot10^{-3}$	72.4 – 113	26.3 – 41

The results show a good correlation with the experimental data obtained previously [6, 10]. However, the mechanisms of development and evolution of a defective network in various conditions require detailed study.

In this paper, as part of a theoretical study, the formation of interstitial atoms and vacancies, as well as their clusters in the form of dislocation loops and voids in CdTe under electron irradiation is modeled.

Using effective activation energy values and effective concentrations of interstitial c_I atoms and c_V vacancies, we can write the equations simulating changes in the concentrations c_I and c_V, as well as the growth of interstitial loops with radii r_I and concentrations c_{bI} and voids with radii r_V and concentrations c_{bV} in the following form:

$$\frac{dc_I}{dt} = c_0 G - K_2 c_I c_V - 2K_I c_I^2 - K_I \frac{2\pi r_1}{b} c_I c_{bI} - K_{SI} c_I c_S,$$
$$\frac{dc_V}{dt} = c_0 G - K_2 c_I c_V - 2K_V c_V^2 - K_V \frac{4\pi r_V^2}{b^2} c_V c_{bV} - K_{SV} c_V c_S,$$
$$\frac{dc_{bI}}{dt} = c_0 K_I c_I^2, \tag{1}$$
$$\frac{dc_{bV}}{dt} = c_0 K_V c_V^2,$$
$$\frac{dr_I}{dt} = c_0 K_I c_I b,$$
$$\frac{dr_V}{dt} = c_0 K_V c_V b.$$

Here $c_0 = 1.5 \cdot 10^{22} cm^{-3}$ is the concentration of the nodes of the unirradiated crystal, c_S is the concentration of "sinks" on the surface. The values $2\pi r_I / b$ and $4\pi r_V^2 / b^2$ are the number of places where point defects attach along the perimeter of the loop and on the void surface, respectively, and b is a parameter of the order of interatomic distance magnitude (Burgers vector). The product of the electron scattering cross section at the nodes of the crystal lattice $\sigma = 3 \cdot 10^{-22}\ cm^2$ and the electron flux density $j = 10^{19} cm^{-2} c^{-1}$ is denoted by $G = \sigma \cdot j = 3 \cdot 10^{-3} s^{-1}$, then the value $G \cdot c_0$ is the rate of generation of point defects (Frenkel pairs). If $\nu = 10^{13} s^{-1}$ is the frequency of atomic oscillations in the lattice, E_2 is the recombination energy of point defects, E_I is the energy of interstitials formation, E_V is the energy of vacancies formation, then the coefficients of the corresponding recombinations and generations can be written as $K_{2,I,V} = b^3 \nu \cdot \exp(-E_{2,I,V} / kT)$, respectively, where k is the Boltzmann constant, and T is the temperature.

As a result of the diffusion of point defects to the surface of a thin sample, they recombine with the surface. The concentration of point defects that change as a result of their migration to the surface can be represented as

$$\left.\frac{\partial c_{I,V}}{\partial t}\right|_{\text{surface}} = K_S c_{I,V} c_S = -D_{I,V} \frac{\partial^2 c_{I,V}}{\partial z^2}, \tag{2}$$

here z is the coordinate perpendicular to the surface of the sample, the values of $D_{I,V} = b^2 \nu \cdot \exp(-E_{mI,mV} / kT)$ are the diffusion coefficients of interstitial atoms and vacancies, and the migration energies of these defects are denoted by E_{mI} and E_{mV}, respectively.

Taking into account the formula (2), it is possible to convert expressions (1) into a system of partial differential equations taking into account the decrease in the number of lattice nodes under irradiation as

$$\dot{u}_0 = -Gu_0 + c_0 K_2 u_1 u_2,$$
$$\dot{u}_1 = Gu_0 + D_I u_1'' - c_0 K_2 u_1 u_2 - 2c_0 K_I u_1^2 - 2\pi c_0 K_I u_1 u_3 u_5,$$
$$\dot{u}_2 = Gu_0 + D_V u_2'' - c_0 K_2 u_1 u_2 - 2c_0 K_V u_2^2 - 4\pi c_0 K_V u_2 u_4 u_6^2,$$
$$\dot{u}_3 = c_0 K_I u_1^2, \tag{3}$$

$$\dot{u}_4 = c_0 K_V u_2^2,$$
$$\dot{u}_5 = c_0 K_I u_1,$$
$$\dot{u}_6 = c_0 K_V u_2,$$

where $u_0 = c(t)/c_0$, and $c(t)$ is the concentration of the lattice nodes during irradiation, $u_1 = c_I/c_0$, $u_2 = c_V/c_0$, $u_3 = c_{bI}/c_0$, $u_4 = c_{bV}/c_0$, $u_5 = r_I/b$, $u_6 = r_b/b$. The dot and dash represent the partial derivatives of the variables t and z, respectively.

The numerical solution of the obtained system of equations gives the dependences $u_0,\dots,u_6$ on z and the irradiation time t at the temperature $T = 300K$ (the parameter values are taken from [6]) (figures 1-4).

The results presented in figure 1 show that in the center of sample $z = 100nm$ the mobility of interstitial atoms is higher than that of vacancies. Interstitial atoms quickly move to the periphery, their concentration decreases faster than the concentration of vacancies. In the time interval from 0.01 to 0.1 seconds, the concentration of vacancies reaches a maximum that almost coincides with the minimum concentration of interstitial atoms. This behavior is explained by the dynamics of the generation and recombination of Frenkel pairs, as well as the agglomeration of Frenkel pairs in the di-interstitials and divacancies, which are supposed to be the nuclei for the formation of dislocation loops and vacancy voids.

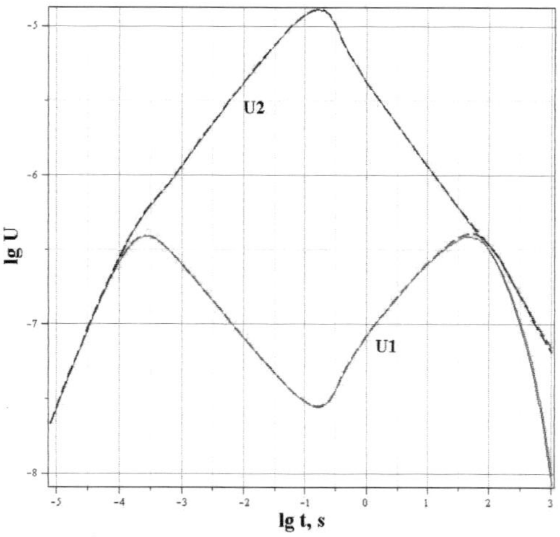

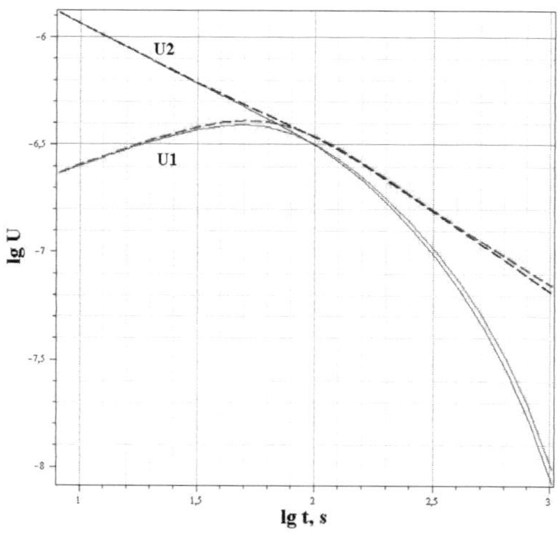

Figure 1. Dependence of u_1 and u_2 on the irradiation time at 300 K. (left, $10^{-5} s \le t \le 10^3 s$, right, $10 s \le t \le 10^3 s$).

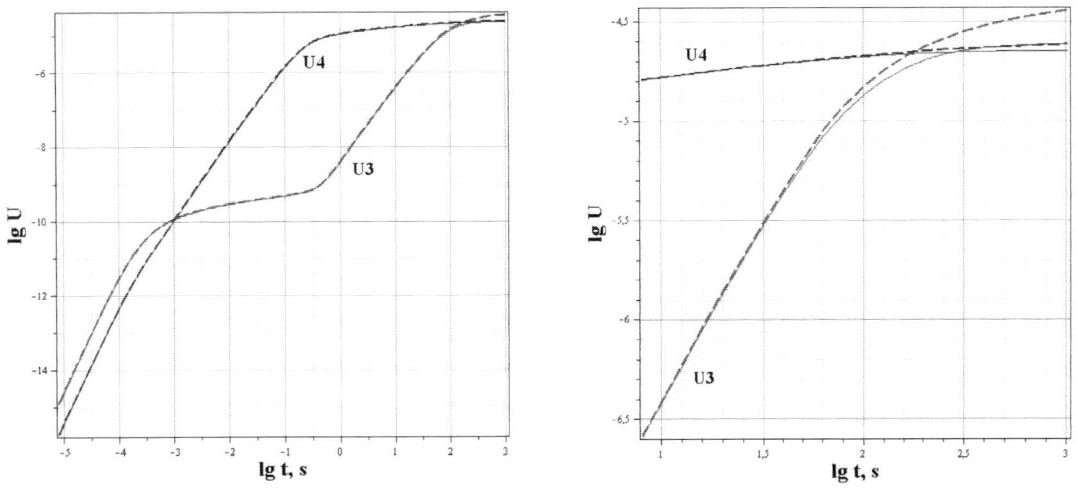

Figure 2. Dependence of u_3 and u_4 on the irradiation time at 300 K. (left, $10^{-5} s \le t \le 10^3 s$, right, $10 s \le t \le 10^3 s$).

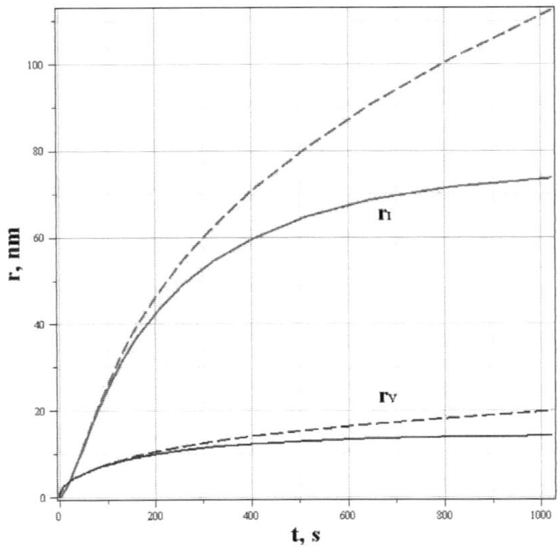

Figure 3. Dependence of the radii of loops and voids on the irradiation time at 300 K.

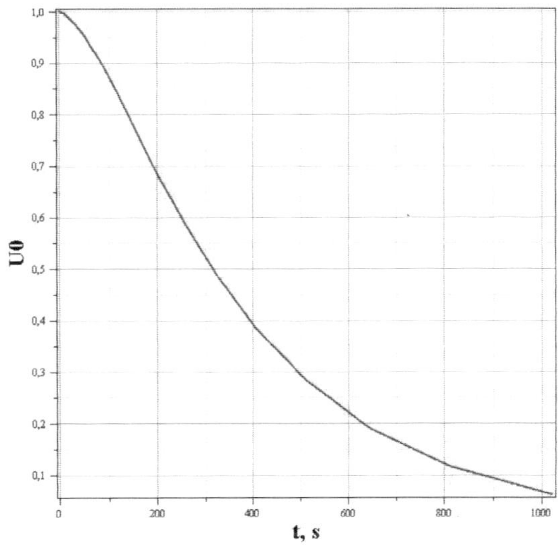

Figure 4. Dependence of the number of lattice nodes on the irradiation time at 300 K.

When the sample is irradiated for more than 50 seconds, both concentrations c_I and c_V decrease. Figures 2 and 3 show that the increase in the concentration of dislocation loops c_{bI} and voids c_{bV}, as well as their radii r_I and r_V, significantly slows down. It should be noted that the real radii of dislocation loops and voids are greater than the lattice parameter. In the figures, these radii begin to exceed the parameter b only a few seconds after the start of irradiation, and their values at long times are almost independent of the initial conditions.

Figure 4 shows the change in the number of lattice nodes during irradiation. A model that takes this change into account is displayed in solid lines in the figures. A model with a constant number of lattice nodes is indicated by dashed lines in the figures. The difference between these models is noticeable in figures 1-3 only at long times.

3. Conclusion

The obtained data were compared with experimental data both for determining the "critical" radii of dislocation loops and for numerical simulation of the formation of point defects clusters of interstitial and vacancy types in cadmium telluride [6, 9]. There is a good agreement between theory and experiment. It is particularly important that we were able to significantly improve the results by taking into account two important features in the process of defect formation in CdTe. First, a contribution from partial Shockley dislocations was added to the model describing the formation of Frank loops. Second, the simulation takes into account the decrease in the number of nodes of the semiconductor crystal lattice with the time of irradiation, which is an important physical point, since the crystal (foil) has a limited size when irradiated *in situ* in TEM.

Taking into account the foregoing, the obtained results quite adequately reflect the mechanisms of defect formation in cadmium telluride and can be applied to similar semiconductors taking into account their physicochemical properties and the value of the SFE.

References

[1] Elsharkawy M, Kanda G, Abdel-Hady E and Keeble D 2016 *Appl. Phys. Lett.* **108** 242102
[2] Veale M, Bell S, Seller P, Wilson M and Kachkanov V 2012 *Journ. of Instrumentation* **7** P07017
[3] Ogawa K and Muraishi M 2010 *IEEE Transactions on Nuclear Science* **57** 17
[4] Limousin O, et al. 2011 *Nuclear Instruments and Methods in Physics Research A* **647** 46
[5] Iniewski K and Awadalla S 2015 *Solid-State radiation detectors: technology and applications* (CRC Press LLC)
[6] Loginov Y, Brown P and Durose K 2003 *The Regularities of Structural Defect Formation in Semiconductors II–VI* (Logos, Moscow)
[7] Gué A, Djafari-Rouhani M, Estéve D and IdrissiSaba H 1991 *J. Phys. I* **1** 97
[8] Loginov Y, Mozzherin A and Paklin N 2019 *IOP Conf. Series: Materials Science and Engineering* **467** 012007
[9] Loginov Yu and Brown P 1992 *Phys.Stat.Sol.(a)* **132** 323
[10] Loginov Y, Mozzherin A and Brilkov A 2014 *Phys. Solid State* **56** 720
[11] Gorichok I 2012 *Phys. Solid State* **54** 1373

Magnetoresistance of rare-earth-substituted garnet ferrite films

A N Masyugin[1], S S Aplesnin[1,2], Y Y Loginov[1] and A I Stognij[3]

[1]Reshetnev Siberian State University of Science and Technology, 31 Krasnoyarskii rabochii prospekt, Krasnoyarsk, 660037, Russia
[2]Kirensky Institute of Physics SB RAS, 50/38 Akademgorodok St., Krasnoyarsk, 660036, Russia
[3]SSPA "Scientific-Practical Materials Research Centre of NAS of Belarus", 19 P. Brovki St., Minsk, 220072, Belarus

E-mail: aplesnin@sibsau.ru

Abstract. Thin epitaxial films of garnet ferrite were studied using impedance spectroscopy. The influence of the magnetic field on the transport characteristics was investigated on the basis of the current-voltage characteristic characteristics (I-V) measured without a field and in a magnetic field H = 12 kOe. An increase in resistance in bismuth ferrite garnet was detected in a magnetic field at a temperature above room temperature.

1. Introduction
The colossal magnetoresistance in inhomogeneous semiconductors with phase separation [1-4], the coexistence of crystallographic phases [5, 6] and polarons is being actively studied due to the possible use of materials in spintronics [7]. In recent years, interest has increased in the study of electrical properties in a magnetic field for paramagnetic materials, which can be used to create random access memory [8-10]. The exchange coupling between localized and delocalized electrons leads to anomalies in the temperature dependence of electrical resistance and formation of magnetoresistance, which is pronounced especially strong in the compounds with the 4f rare-earth elements [11]. In semiconductors, the spin–orbit interaction at the interface can attain high values and, according to the theoretical calculations [12–13], leads to a significant increase in the magnetoresistance at the heating in the paramagnetic state. In magnetic films, a large potential gradient at the interface can induce a spin-orbit interaction.

A linear magnetoelectric effect [14–15] and the coexistence of current channels with ferroelectric regions were found in bismuth ferrite garnet films, which makes it possible to control the current with a magnetic and electric field.

Purpose of the study is to establish the influence of a magnetic field and its mechanism on the transport characteristics of garnet ferrite in results of the replacement of yttrium with bismuth and cerium ions.

2. Experimental results and discussion
Iron garnets are good insulators with high Curie temperature. In the deposition of films with a thickness of 100-200 nm, the film – substrate interface plays a significant role, where the formation of charged defects and the polycrystalline structure of the films are possible. In this case, the formation of magnetic polarons at the interface is possible. Spin-phonon interaction can change the magnetic structure [16-19].

Garnet ferrites with cerium exhibit charge fluctuations of Ce +3- Fe+3=Ce+4-Fe+2 with a gap of 1.3 eV. At the interface and at the polycrystalline boundaries, fluctuations in the crystal field will lead to a decrease in the gap. Oxygen vacancies can also lead to ferrous iron with the formation of electron-hole states. Perhaps two transport channels along the anionic and cationic systems. This can be established by the method of impedance and IR spectroscopy, from the current-voltage characteristic and electrical resistance in magnetic fields up to 12 kOe.

We studied epitaxial films $Nd_1Bi_2Fe_5O_{12}$(450nm)/$Nd_2Bi_1Fe_4Ga_1O_{12}$(90nm) on a glass substrate and $Nd_{0.5}Bi_{2.5}Fe_5O_{12}$(450nm) on a GGG single crystal substrate grown in the (111) crystallographic direction. The films have a ferrimagnetic structure with a Curie temperature Tc = 450 K. In the vicinity of Tc, the conductivity and impedance exhibit a kink characteristic of the polaron type of current carriers. This indicates the exchange interaction of the spins of mobile electrons with localized spins.

The influence of the magnetic field, the transport characteristics were studied on the basis of the I-V measured without a field and in a magnetic field H = 12 kOe applied perpendicular to the film and along the film. Figure 1 shows the I-V of films on a glass and GGG substrate. The I-V hysteresis is associated with the presence of regions with electric polarization and is shifted along the axis of the electric field by 20–30 V for the film on GGG. When heated above 280 K, the hysteresis width decreases several times and the displacement field disappears. In a magnetic field, the resistance decreases below room temperature and increases above 280 K. The magnetoresistance $\Delta R=(R(H)-R(0))/R(0)$ exhibits anisotropy relative to direction of the magnetic field to the normal of the film. This is due to the orientation of the magnetization of the film, which reveals a spin-reorientation transition at the room temperature. The I-V of the film on the glass (figure 1 b) has a rectangular shape and the current does not depend on voltage up to 320 K. In this film, electric polarization is induced by an external electric field and a pyrocurrent Ip = dP / dt is arised, the value of which exceeds the external current. At the heating the resistance of the film decreases and the external current exceeds the Ip current. In a magnetic field, the hysteresis loop practically does not change. An increase in the resistance in a magnetic field (R(H)-R(0)) /R(0) ~ 0.01 was found on alternating current up to 300 K. The dependence of ΔR (H) on the field is satisfactorily described by the power-law function ΔR (H) ~ H^2 (figure 2a).

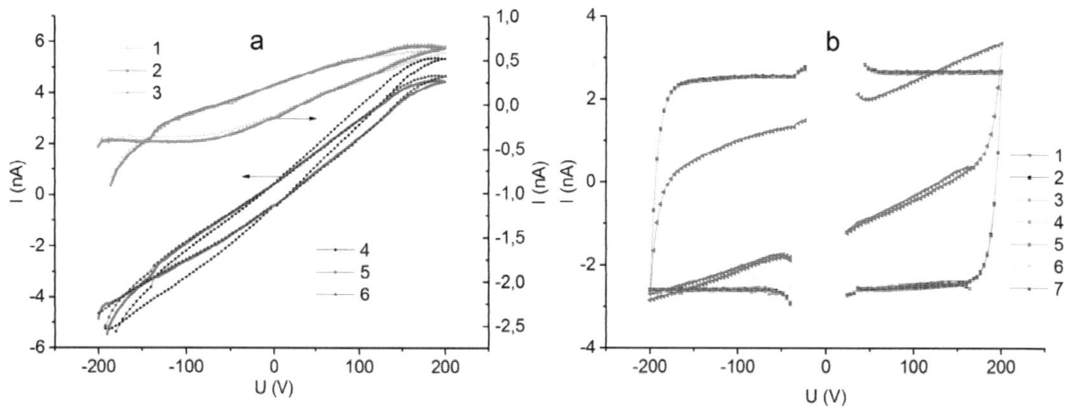

Figure 1. (a) - I-V of films on a GGG substrate at 160K H = 0 (1), H = 12kOe perpendicular to the film (2), along the film (3), at 280K H = 0 (4), H = 12kOe perpendicular to the film (5) along the film (6). (b) I-V of films on a glass substrate at 380K H = 0 (1), 120K H = 0 (2), H = 12 kOe directed at angles 0 ° (3), 90 ° (4), 180 ° (5), 270 ° (6), 360 ° (7) relative to the normal to the film.

The resistance in the frequency range 100-500 Hz increases in the magnetic field by 5-8% at T = 400 K and the magnetoresistance ΔR = (R (H) -R (0)) / R (0) is 2-3% at T = 350 K and 450 K, and disappears at room temperature. Magnetoimpedance does not exceed 0.3%.

At these temperatures, the activation energy decreases from 0.27 eB to 0.18 eB. DC resistance increases smoothly with heating and reaches a maximum at T = 300 K with a further decrease. In a magnetic field, the change in current does not exceed one percent.

Domain and crystal boundaries can be determined from impedance. In the analysis of the experimental results, the equivalent circuit approximation was used. The impedance hodograph is described by one semicircle, corresponding to the R, C circuit, where the capacitance is due to crystalline boundaries (figure 2b). In the high frequency region, a linear contribution to the impedance is added. This element models the impedance of a linear diffusion process occurring in a homogeneous layer with a finite thickness. The impedance in this process depends on the resistance, particle diffusion coefficient, and the thickness of the Nernst diffusion layer. At high frequencies, the experimental data are well approximated by a linear Warburg impedance diagram [20,21]:

$$Z_W(j\omega) = k_f(j\omega D_0)^{-\frac{1}{2}} = k_f(\omega D_0)^{-\frac{1}{2}}(1-j) \tag{1}$$

where k is the coefficient, D is diffusion. The diffusion contribution to the impedance is observed at room temperature and above. Diffusion decreases with increasing frequency. The impedance hodograph shown in figure 2b is described in the model of equivalent circuits by one semicircle. The relaxation time τ = RC (R-resistance, C-capacitance) drops sharply when heated from τ = 10 s at T = 300 K to τ = 0.1 s at T = 340 K. An increase in the relaxation time with decreasing temperature indicates the formation of a charge glass state at room temperature. Broken Bi-O bonds on the interface form a dipole moment and regions with local electric polarization. This polarization is shielded by free charge carriers formed by impurities and structural defects.

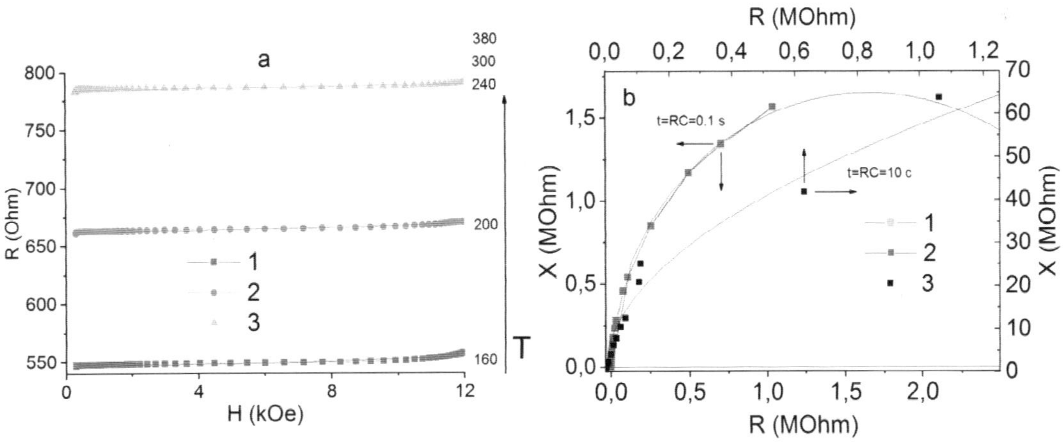

Figure 2. (a) – Dependence of the active resistance of a film on a glass substrate on the magnitude of the magnetic field at temperatures of 160K (1), 240K (2), 380K (3), (b) —impedance hodographs at temperatures of 340K H = 0 (1), H = 12 kOe (2), 300K H = 0 (3).

The participation of electrons of a alone bismuth pair in hybridization is confirmed by the presence of a magnetic field on Bi nuclei. A change in the length of the cation-anion bond leads to a displacement of oxygen and to a change in the exchange interaction between oxygen ions and to uniaxial anisotropy. An external magnetic field can switch the direction of magnetic anisotropy and the electric polarization of the Bi-O bond. In a spin-reorientation transition, the mobility of current carriers in an external magnetic field changes as a result of s-d interaction. This mechanism explains the change in the sign of magnetoresistance in temperature. The scattering of current carriers by magnetic inhomogeneities, the magnetic moments of which can be changed by a magnetic field, was previously considered on sulfides [22-24].

$Ce_3Fe_5O_{12}$ films were obtained on a GGG single crystal substrate grown in the (111) crystallographic direction by thick 200 nm. Two transitions were found from the optical absorption spectra associated with the transition of electrons between the $^6A_1 \rightarrow ^4T_2$ levels in the tetrahedron at 2.3 eV and with the charge transfer 2p→3d through oxygen Fe_{tetr}–O–Fe_{oct} at 2.6 eV. «Tails» in the range 1.7–2.3 eV are due to fluctuations in the crystal field in tetrahedra at the interface and grain boundaries. The energy of charged defects are determined from the IR spectra. The IR spectrum contains two absorption lines in the frequency range ω1 ~ 5520 cm^{-1} and ω2 = 6800 cm^{-1}. One line at ω1 disappears at 300 K, the other at 400 K.

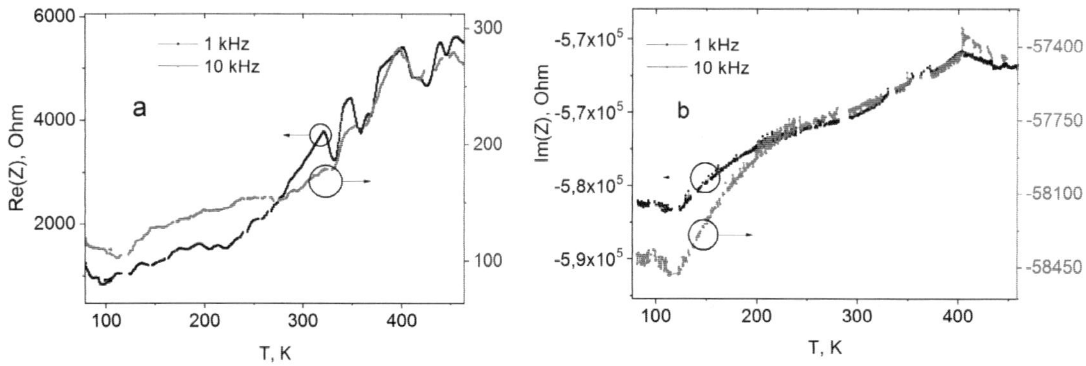

Figure 3. Real (a) and imaginary (b) parts of the Ce3Fe5O12 impedance at frequencies ω = 1 kHz, 10 kHz versus temperature.

The disappearance temperature of the second line coincides with the maximum temperature of 400 K of the imaginary part of the impedance (figure 3a), which is independent of frequency. The maximum dielectric constant is caused by charge fluctuations, which also lead to a local maximum of resistance on alternating current (figure 3a). The resistance in the frequency range 100-500 Hz increases in the magnetic field by 5-8% at T = 400 K and the magnetoresistance ΔR=(R(H)-R(0))/R(0) is 2-3% at T = 350 K and 450 K, and disappears at room temperature.

3. Conclusion

In bismuth ferrite, an increase in resistance in a magnetic field was detected at temperatures above room temperature. The maximum magnetoresistance of the order of several percent is achieved at a temperature of 450 K. The impedance hodograph is described by a single circuit and at high temperatures the diffusion contribution is added. Magnetic impedance not exceeding one percent was found.

References

[1] Zhu Y, Du K, Niu J, Lin L, W Wei W, Liu H, Lin H, Zhang K, Yang T, Kou Y, Shao J, Gao X, Xu X, Wu X, Dong S, Yin L and Shen J 2016 *Nat. Commun* **7** 1250
[2] Tokura Y 2006 *Reports Prog. Phys.* **69** 797
[3] Aplesnin S S, Ryabinkina L I, Romanova O B, Sokolov V V, Pichugin A Y, Galyas A I, Demidenko O F, Makovetski G I and Yanushkevich K I 2009 *Phys. Solid State* **51(4)** 698
[4] Aplesnin S, Romanova O, Har'kov A, Balaev D, Gorev M, Vorotinov A, Sokolov V, and Pichugin A 2012 *Phys. Status Solidi B.* **249** 812
[5] Bebenin N G, Zainullina R I and Ustinov V V 2018 *UFN* **188** 801
[6] Udod L V, Aplesnin S S, Sitnikov M N and Molokeev M S 2014 *Phys. Solid State* **56(7)** 1315
[7] Fert A 2008 *UFN* **178** 1336
[8] Aplesnin S S and Sitnikov M N 2014 *JETP Letters* **100(2)** 95-101
[9] Aplesnin S S, Romanova O B and Yanushkevich K I 2015 *Phys. Status Solidi B* **252(8)** 1792-1798

[10] Aplesnin S S, Ryabinkina L I, Romanova O B, Velikanov D A, Balaev A D, Balaev D A, Yanushkevich K I, Galyas A I, Demidenko O F and Bandurina O N 2008 *J. Exp. Theor. Phys.* **106(4)** 765-772
[11] Olson Reichhardt C J, Reichhardt C and Bishop A R 2004 *Phys. Rev. Lett.* **92** 4
[12] He H T, Liu H C, Li B K, Guo X, Xu Z J, Xie M H and Wang J N 2013 *Appl. Phys. Lett.* **103** 031606.
[13] Sánchez J C R, Vila L, Desfonds G, Gambarelli S, Attané J P, De Teresa J M, Magén C and Fert A 2013 *Nat. Commun.* **4** 2944
[14] Aplesnin S S, Masyugin A N, Sitnicov M N, Rybina U I and Ishibashi T 2018 *J. Magn. Magn. Mater.* **464** 44
[15] Popova E, Shengelaya A, Daraselia D, Japaridze D, Cherifi-Hertel S, Bocher L, Gloter A, Stéphan O, Dumont Y and Keller N 2017 *Appl. Phys. Lett.* **110** 142404
[16] Aplesnin S S 1996 *Phys. Solid State* **38** 1031
[17] Aplesnin S S 1998 *J. Phys. Condens. Matter* **10** 10061
[18] Aplesnin S 2000 *Phys. Rev. B.* **61** 6780
[19] Aplesnin S S 2003 *J. Exp. Theor. Phys.* **97** 969
[20] Scherson D A and Newman J. J. 1980 *Electrochem. Soc.* **127** 110
[21] Huang J 2018 *Electrochim. Acta* **281** 170
[22] Petrakovskii G A, Loseva G V, Ryabinkina L I and Aplesnin S S 1995 *J. Magn. Magn. Mater.* **140–144** 147
[23] Aplesnin S S, Romanova O. B., Gorev M V, Velikanov D A, Gamzatov A G and Aliev A M 2013 *J. Phys. Condens. Matter* **25**
[24] Aplesnin S S, Ryabinkina L I, Romanova O B, Velikanov D A, Balaev A D, Balaev D A, Yanushkevich K I, Galyas A I, O. F. Demidenko O F and Bandurina O N 2008 *J. Exp. Theor. Phys.* **106** 765

The dynamics of pore formation in the electron beam welding

T G Oreshenko and O A Bocharova

Reshetnev Siberian State University of Science and Technology, 31 Krasnoyarskii rabochii prospekt, Krasnoyarsk, 660037, Russia

E-mail: veisver@mail.ru

Abstract. The main mechanisms of porosity formation in electron beam welding are described by authors. The dynamics of changes in the shape, location and velocity of the pores in the weld pool is considered. The authors present the relationships linking the parameters of the electron beam welding process with the shape and location of the pores and their experimental data.

1. Introduction

The pore formation is a complex of the physical and chemical phenomena. Its development proceeds in several stages: the formation of micropores, the development of pores in the volume of the weld pool during diffusion and coalescence, the degassing of the bath due to the emergence of bubbles [1]. The industrial aluminum has already had micropores. Therefore, the formation of pores in aluminum is considered as a micropore development process in pores. Micropores appear in the interdendritic space, at the locations of inclusions, impurities, alloying elements. During crystallization process some bubbles do not have time to leave the weld pool turn into pores. An increase in micropores in the volume occurs in a liquid metal as a result of diffusion of hydrogen dissolved in the metal in these structures. Atomic hydrogen in bubbles transforms into a molecular form and creates a certain pressure in them, balanced by the external pressure exerted on the bubble. When the pore size is small, it will remain spherical, due to the surface tension force. Also when the pores grow in the inter dendritic space, the pore's shape will noticeably differ from spherical due to the non-simultaneous solidification of the base metal and inclusions [2].

2. Materials and research techniques

The physics of the process of electron beam welding (EBW) is considered in more detail in order to creation a model of pore formation.

A single pore is in the liquid metal of the bath. The further behavior of the pore will be determined by the buoyancy (Archimedean) force and the resultant surface tension forces in a nonuniformly heated liquid. In addition, the pore will move along with the movement of the weld pool metal, the nature of which depends on the shape and size of the bath, the location of the bubble and other factors. In most scientific sources, the dynamics of the ascent of a bubble is considered without taking into account unsteady bonds, but it has an important role in a number of many processes. It is necessary to evaluate the dynamics of the change in the shape of the bubble in various modes of ascent with considering the unsteady forces.

For determining the emergence of a spherical pore dynamics with a diameter D in the fluid, that is, to find the dependence of the bubble ascent on time $u(t)$, we can use the equation of motion of the bubble in the form taking into account gravity of Newton's second law, Archimedes force, fluid viscosity and Stokes mode of motion in the projection onto the z axis directed vertically upward (opposite to the direction of the gravitational acceleration vector g)

$$\frac{du}{dt} = g(\frac{\rho_{lic}}{\rho_g} - 1) - \frac{18\mu}{\rho_g D^2} u \tag{1}$$

where ρ_{lic} is the density of the melt, ρ_g is the density of the gas, is the dynamic viscosity of the liquid.
For the case of stationary ascent

$$u_0 = \frac{(\rho_{lic} - \rho_g) D^2}{18\mu} g \tag{2}$$

The stationary ascent velocity of the bubble is proportional to the square of the diameter. The maximum bubble diameter $D_{\max}$, for which the condition of the Stokes regime is satisfied

$$D_{\max} = \sqrt[3]{\frac{18\mu^2}{\rho_{lic}(\rho_{lic} - \rho_g) g}} \tag{3}$$

Other dynamic parameters are determined by solving equation (1). So, dependence of the bubble ascent rate on time

$$u(t) = u_0 \left[1 - \exp\left(-t \frac{18\mu}{\rho_g D^2} \right) \right] \tag{4}$$

Ascent of the bubble with regard to the attached mass is slower than without the attached mass.

Since the temperature in the weld pool is not distributed evenly, the moving bubble will fall into zones with different temperatures, which will lead to a change in its ascent rate, since the density and viscosity values included in the formula depend on the melt temperature

Because the temperature in the weld pool is not distributed evenly, the moving bubble will fall into zones with different temperatures. It will lead to a change in its ascent rate, i.e. density and viscosity values included in the formula depend on the melt temperature.

$$\left. \begin{aligned} \rho_g &= a - b(T - T_m) \\ \mu &= c + \frac{d}{T^2} \\ \sigma &= \sigma_m - \frac{d\sigma}{dT}(T - T_m) \end{aligned} \right\} \tag{5}$$

where $a, b, c, d, \frac{d\sigma}{dT}$ - coefficients that depend on the type of metal or alloy; T_m, σ_m - temperature and surface tension at the melting point, K.

The resultant of surface tension is the result of uneven compression of the bubble. This force is directed towards increasing the temperature of the metal and tends to move the gas bubble almost normal. The portable velocity of the gas bubble in the bath is equal to the velocity of the liquid metal, the value of which depends on the parameters of the welding mode. The axial flow rate of the molten metal reaches its maximum value in the lower part of the bath crater and slows down as it moves to the caudal part of the bath.

The conditions for further growth of the pore surface F for EBW can be obtained from the model described in [3]. If we neglect the pressure in the welding chamber

$$F > 5,4\frac{\sigma^3}{P_v} \tag{6}$$

If the inequality is not satisfied, then the gas from the bubble dissolves in the metal. The obtained relations should be supplemented by temperature and viscosity distributions during electric welding

$$T = T_0 + \frac{Q}{2\pi\lambda\delta}e^{-\frac{vx}{2a}}\left(r\sqrt{\frac{v^2}{4a^2}+\frac{bc\rho}{\lambda}}\right) \tag{7}$$

where ρ – melt density, c – constant pressure heat capacity, T – absolute temperature, v– welding speed, Q – heat from available sources, λ – coefficient of thermal conductivity, x, r – coordinates, δ – thickness of welded parts, b – alloy dependent coefficient.

Substituting value T in σ, and then, in the inequality under consideration, we obtain

$$F > 5,4\frac{\left(\sigma_0 - \frac{KQre^{-cvx\rho}\sqrt{c^2\rho^2v^2+4b\lambda\rho}}{4\pi\lambda(\sigma_0 - K(T-T_0))}\right)^3}{P_v} \tag{8}$$

The modulation of the beam current is not considered by us. Due to the fact that many attempts beam current modulation, as well as the oscillation level of the focus when welding metals of medium and large thickness does not give positive results (there are intense metal splashing and the formation of undercuts on both beam sides) [4].

Nevertheless, when scanning the joint of the product with an electron beam, the temperature distribution changes on the surface of the parts being welded, centrifugal forces arise. The temperature field on the surface in the circular penetration channel scan can be represented as follows [5]:

$$T \approx T_a + \Delta T\sin(2\pi ft - \varphi), \Delta T \approx 0.4\frac{q\eta\sin\alpha}{\lambda}\sqrt{\frac{ad}{fr}}, \tag{9}$$

where φ – azimuthal angle between the direction of welding and the considered point of the weld pool, T_a – average temperature of the vapor-gas channel, ΔT – amplitude of temperature change, q – power density in an electron beam (in thermal units), η – beam transfer efficiency, d – beam diameter, α – the angle of the beam in the surface of the penetration channel.

The sinusoidal temperature component causes turbulent flow in the molten pool and a transverse force component acting on the gas bubbles, causing the rise of a depth along a helical path with a radius equal to the radius of the scan plus the width of the heat affected zone.

Under the conditions of equality of all forces (Basset, Archimedes, gravity, surface tension and the force arising under the action of scanning an electron beam), the shape of the bubble will be close to spherical. The predominance of one of them will change shape in a certain direction. A joint solution of equations (4)-(9) allows us to predict the shape of the pores in EBW.

3. Results and Discussion

Increasing in the welding speed to a certain value, the number of pores in the weld significantly decreases. This is due to the decrease of the temperature influence on the metal parts and as a consequence, decreasing the transition of pore nuclei to larger pores. Increasing speed of pores become smaller, their shape is spherical. When the critical speed is exceeded, pores again begin to appear in the weld in the region where the gas-vapor channel passes through due to rapid cooling (figure 1).

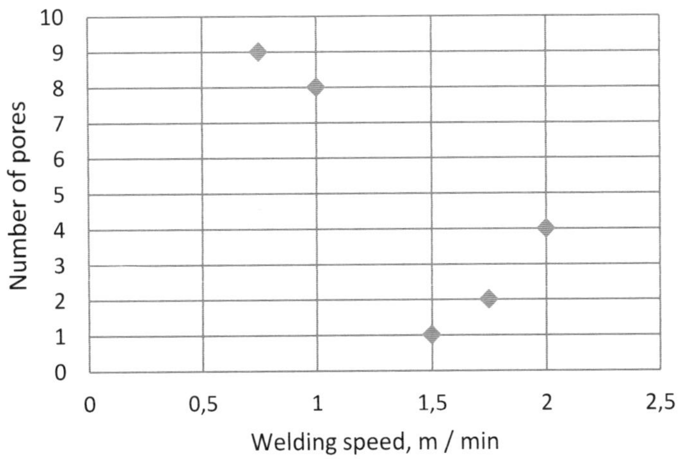

Figure 1. Dependence of the number of pores on the welding speed of an aluminum-titanium alloy.

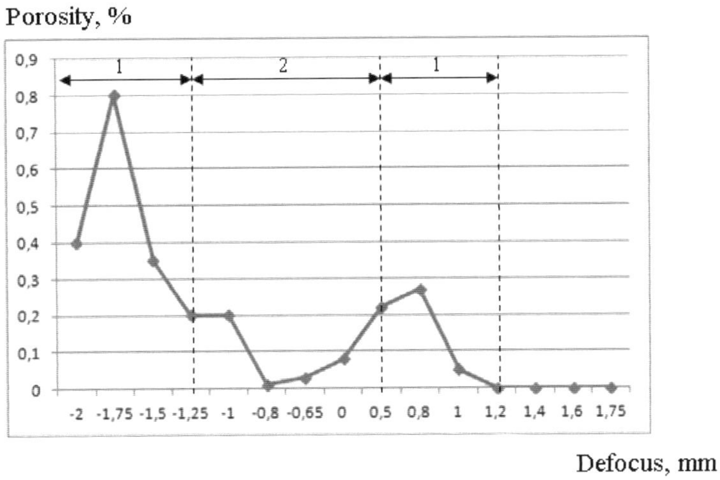

Figure 2. Porosity formed at various values of beam defocusing at a welding speed of 0.5 m / min: 1 - unstable vapor-gas channel; 2 - stable vapor-gas channel.

The pores are in different parts of the joint as in the upper part and closer to the root. The gas bubbles have time to rise closer to the surface and pores located above at lower velocities. At high welding speeds, the metal cools and hardens faster, preventing the pores from growing and rising higher.

At intermediate welding and cooling speeds, so-called wormholes are formed the interdendritic pores that have an elongated shape. Their shape is due to the growth of the pore nucleus in the intergranular space, when most of the crystal has already formed.

Porosity is formed not only in the liquid state, but also in the solid phase. For example, in the heat affected zone, or in the hardened weld pool during refrigeration process.

Such porosity has a size of about 1-2 microns in diameter. However, in multi-pass welding, such pores are combined and can grow. Using of preliminary passages in order to heat the weld pool contributes to better degassing of the weld pool due to the fact that, during preliminary pass, the pore nuclei are combined into larger bubbles. When the electric welding performs, these bubbles will have a sufficient diameter for quick exit to the surface.

Also defocusing of the electron beam has a significant effect on the formation of porosity in an electric welding. This value determines the porosity caused collapsed vapor-gas channel. Studies have shown that beam focusing has a selective effect on the amount of porosity (figure 2).

With a decrease in the welding speed of the 5654 alloy from 50 to 10 m / h, the pore volume increases by more than 40 times. It well correlates with early studies in this direction [6]. The pore content increases with a decrease in speed below the critical one - for 5654 this is 10 m / h. This can be explained by overheating of the weld pool and intense evaporation of magnesium. Increasing speed of the amount of welding gas pores in the welded joints of the alloy 5654 significantly reduced. The density of stitches is increased. In this case, the pore has a spherical shape.

4. Conclusion

A change in the shape of the pores and the speed of their movement at the junction indicates the dependence of the shape on the technological parameters of the welding process. Increasing spot heating zone and temperature influence the quantity and pore's size increase. At welding speeds below critical, narrow elongated wormhole pores are formed. This is due to the accumulation of gas between the crystallites of the solidifying metal. Based on this, technological recommendations can be developed aimed at reducing porosity in the process of welding of experimental and developed alloys.

References

[1] Zusin V Ya and Serenko V A 2004 *Welding and surfacing of aluminum and its alloys* (Mariupol: Renata) p 468

[2] Legait P-A 2005 *Formation and distribution of porosity in Al-Si welds* : A thesis submitted to the faculty of the Worcester polytechnic institute in partial fulfillment of the requirements for the degree of masters of science in material science and engineering (Worcester) p 110

[3] Hakobyants K S 1984 Prevention of the formation of root defects during electron beam welding *Automatic welding* (Kiev) **6** 375 pp 59-61

[4] Kaydalov A A and Nazarenko O K 1986 The main technological methods of electron beam welding *Automatic welding* (Kiev) **4** 397 pp 51-58

[5] Negoda E N 2008 *Thermal processes in welding: textbook. allowance* (Vladivostok: DVGTU) p 125

[6] Bondarev A A, Voropay N M, Ischenko A Ya and Rabkin D M 1972 The influence of technological factors on the porosity of welds on aluminum made by electron beam welding *Automatic welding* (Kiev) **8** 233 pp 24-26

IOP Conf. Series: Materials Science and Engineering 1230 (2022) 012016 doi:10.1088/1757-899X/1230/1/012016

Concentration dependences of the structural, magnetic and kinetic properties of manganese chalcogenides doped with tellurium

O B Romanova[1,2], M N Sitnikov[1], Yu Yu Loginov[1] and Abdelbaki Hichem[1]

[1]Reshetnev Siberian State University of Science and Technology, 31 Krasnoyarskii rabochii prospekt, Krasnoyarsk, 660037, Russia
[2]Kirensky Institute of Physics SB RAS, 50/38 Akademgorodok St., Krasnoyarsk, 660036, Russia

E-mail: rob@iph.krasn.ru

Abstract. The effect of anionic substitution on the structural, magnetic, and kinetic properties of chalcogenides of the $MnSe_{1-X}Te_X$ system in the temperature range of 77-400K in magnetic fields up to 12 kOe was studied. With an increase in the substitution concentration, changes in the structural and magnetic characteristics, which correlate with changes in the electronic structure, were found. The type and mobility of the main current carriers were determined from the Hall effect measurements.

1. Introduction

Magnetic semiconductors undergoing electronic [1–4] and magnetic phase transformations [5–8], both in temperature and in concentration, are promising materials for the development and creation of an element base of microelectronics. The interaction of the magnetic and elastic subsystems leads to a change in the magnetic structure, to new magnetic phases in which the electronic structure and kinetic properties can change [9-11]. Such substances include $MnSe_{1-X}Te_X$ solid solutions synthesized on basis of manganese monoselenide. These compounds show a polaron conduction mechanism, which is sensitive to the slightest changes in the electronic structure, as well as in the unit cells of the magnetic and crystalline subsystems.

Manganese chalcogenides MnSe and MnTe reveal structural, magnetic transitions and a change in transport properties from semiconductor to metallic at temperatures below 300 K. The initial compound manganese monoselenide (MnSe) is an antiferromagnet with T_N=135K [12-14]. In the temperature range 248 K $<T<$ 266 K, MnSe exhibits a structural phase transition from the cubic phase of the NaCl type to NiAs, and below this temperature, these two phases coexist in the sample [15]. MnTe crystallizes in the hexagonal type of NiAs structure and is an antiferromagnetic with a Neel temperature of 310 K. MnSe and MnTe are p-type semiconductors with an energy gap in the spectrum of single-particle electronic excitations for MnSe equal to (2.0-2.5 eV), and for MnTe (0.9-1.3 eV), respectively [16, 17]. In MnSe the magnetoresistance effect was detected in the range Néel temperature, while there is no effect in manganese monotelluride and it is observed only in thin-film samples [18, 19].

The aim of this work is to study the effect of anionic substitution on the structural, magnetic and kinetic properties of chalcogenides $MnSe_{1-X}Te_X$ ($0 \leq X \leq 0.4$) solid solutions.

2. Results and Discussion

Anion-substituted $MnSe_{1-X}Te_X$ solid solutions with a substitution concentration (0 <X <0.4) were synthesized by the solid-phase reaction method using a stepwise mode. The synthesis technique is described in detail in [12].

X-ray diffraction pattern of the powders of the studied compositions were obtained at T = 300 K on DRON-3 X-ray diffractometer (CuK_α radiation). X-ray data show that anionic substitution in the $MnSe_{1-X}Te_X$ system with an increase in tellurium content leads to a noticeable decrease of the intensity of reflexes. The decrease in the intensity of diffraction reflections in X-ray diffraction pattern is due to the fact that the formation of solid solutions at substitution in the anion sublattice, is accompanied by significant crystalline distortions and increase in the unit cell parameter *a* (table 1). The growth of a disorder in the whole cubic structure with a spatial group *Fm3m* (225) is observed. An increase in the cell parameter with growing tellurium concentration allows us to draw a conclusion about the formation of solid solutions of the $MnSe_{1-X}Te_X$ system.

In order to establish the effect of anion substitution on the magnetic subsystem and the possibility of the formation of magnetically inhomogeneous states in the initial MnSe matrix, the magnetic properties of the synthesized samples of the $MnSe_{1-X}Te_X$ system (0<X<0.4) were studied using the Faraday method in a magnetic field of 8.6 kOe in the temperature range 80- 1000K on samples placed in evacuated quartz ampoules. Table 1 presents the basic magnetic characteristics of the synthesized samples of the $MnSe_{1-X}Te_X$ system. With an increase in the substitution concentration, there is a significant decrease in the temperature of the magnetic phase transition (T_N) and paramagnetic Curie temperature (Θ_p) in absolute value, which indicates an increase in ferromagnetic exchange. This assumption confirms the hysteresis of the magnetic susceptibility found in the temperature range 80–350 K for all samples, but it is most pronounced for the composition X = 0.2. With increasing concentration, the magnetic moment decreases from 5,50 μ_B for X=0.1 to ~5,13 μ_B for X=0.4. Substitution of selenium by tellurium leads to an anomalous behavior of the susceptibility in the vicinity temperature of the magnetic phase transition ~330K for pure MnTe. This behavior of the susceptibility is probably due to the formation of inhomogeneous states (nanoclusters) of the MnTe compound in the matrix of the $MnSe_{1-X}Te_X$ solid solution.

Table 1. Magnetic characteristics of the $MnSe_{1-X}Te_X$ system.

Состав, X	0.1	0.2	0.3	0.4
a, nm	0.548	0.554	0.558	0.562
T_N, K	135	130	125	120
Q_p, K	-350	-330	-300	-275

Changes in the structural and magnetic characteristics correlate with changes in the electronic structure with increasing substitution concentration. A similar correlation of changes in the electronic and crystalline structures was observed in bismuth pyrovskites [20, 21]. The electrical resistivity was measured by the standard four-probe direct current method in the temperature range 77–300 K in a magnetic field of up to 12 kOe and indicates the semiconductor type of conductivity of the synthesized compounds of the $MnSe_{1-X}Te_X$ system. A change in the substitution concentration leads to a decrease in the electrical resistivity (figure 1a), which correlates with the data of magnetic properties. In this case, the activation energy $\Delta E \approx$ (0.07-0.09) eV is practically independent of the composition of this system samples.

The influence of the magnetic field on the transport characteristics with a change in concentration was determined from the current-voltage characteristics measured in a zero magnetic field and in a field of H = 10 kOe. The experiment showed that the obtained dependences U (I) are linear and do not depend on the magnitude of the applied magnetic field at temperatures below 100K. It was established that in the magnetic field the resistance of the samples decreases and the greatest change is found in the vicinity of the Néel temperature for X = 0.1, it is 100% (figure 1b). With an increase in the substitution concentration, the magnitude of the magnetoresistance (δ_H) decreases in absolute value and δ_H ~ 5% for the composition X = 0.2 is observed in the paramagnetic area. This value is much smaller in comparison with magnetoresistance in solid solutions based on manganese sulfide [22-24]. A further increase in concentrations to X = 0.4 leads to the disappearance of the effect. One possible explanation for the decrease in resistance in a magnetic field for composition X = 0.1 is an increase in the electron localization radius in potential wells as a result of competition between ferromagnetic and antiferromagnetic interactions and a reduction in the width of the potential barrier.

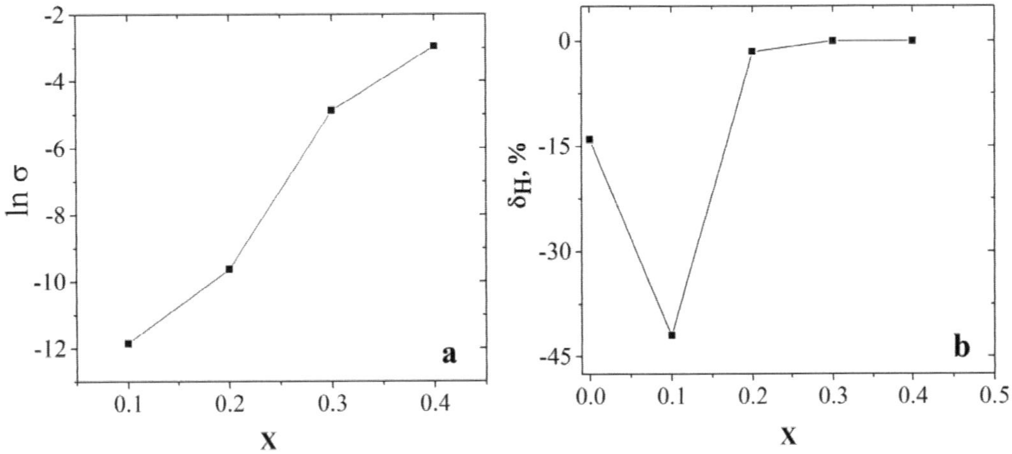

Figure 1. Concentration dependences of the conductivity (a) and the relative magnitude of the magnetoresistance (b) at T = 130K for the $MnSe_{1-X}Te_X$ system.

To determine the sign and mobility (μ) of the current carriers, the Hall effect was measured. The galvanomagnetic properties of anion-substituted $MnSe_{1-X}Te_X$ solid solutions were studied in magnetic fields up to 12 kOe in the temperature range 77–400K.

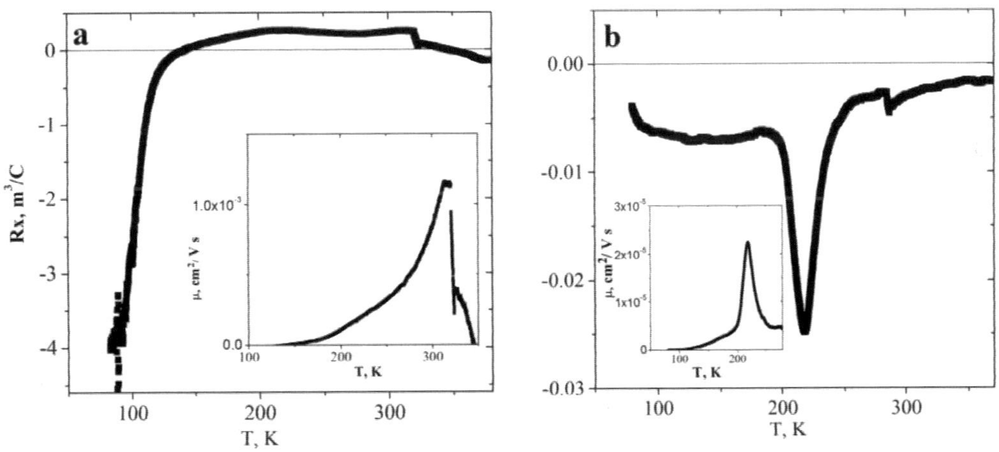

Figure 2. Temperature dependences of the Hall constant for concentrations X = 0.1 (a) and 0.3 (b). The inset shows the temperature dependence of mobility for these concentrations.

The manganese monoselenide MnSe has a p-type conductivity. For small substitution concentrations X = 0.1, with increasing temperature (figure 2a), the Hall coefficient (R_H) increases and at a temperature of 140K the sign of the Hall constant changes in the field where the maximum magnetoresistive effect is observed [25]. In the temperature range 140-347K, the hole type of conductivity dominates with a sharp decrease in the R_H at T = 320K. An increase in the anion substitution of selenium with tellurium in the $MnSe_{1-X}Te_X$ system leads to a change in the sign of current carriers. For the concentration X = 0.3, the dependence R_H (T) shows negative values, which indicates the electronic type of the carrier type (n-type) in the entire temperature range of measurements in fields up to H ~ 12 kOe. In the temperature range 190–250K, the R_H(T) dependence has an anomaly, which is due to the presence of charge carriers of different signs in the solid solutions of the $MnSe_{1-X}Te_X$ system, given that the initial compound is p-type semiconductors, as well as an increase or decrease in their number at different temperatures and with the mobility of charge carriers. The presence of an anomaly on R_H (T) correlates with the magnetization hysteresis observed in this temperature range. It should be noted that in MnSe a sharp maximum of magnetic susceptibility is observed in the temperature range T = 160K - 180K, exceeding the Néel temperature in the cubic modification. In the same temperature range, upon cooling and heating of the MnSe single crystal in the temperature behavior of the resistance, hysteresis is observed, which corresponds to the coexistence of cubic and hexagonal modifications [16, 26]. Moreover, for MnSe, the resistance hysteresis has the form of a loop (as infinity), which depends on the rate of change of the measurement temperature. The mobility of the main current carriers is determined from the Hall effect measurements. It was found that the $MnSe_{1-X}Te_X$ system is characterized by small mobility values, which decrease with increasing X.

3. Conclusion

It has been found that anionic substitution of selenium by tellurium leads to an increase in the unit cell parameter of the fcc lattice of the NaCl type of antiferromagnetic samples of the $MnSe_{1-X}Te_X$ system in the space group $Fm\overline{3}m$ (225). It was determined that an increase in the concentration (X) of substitution leads to a decrease in the Neel temperature and magnetic moment within 20% and the magnitude of the magnetoresistive effect. The value of the energy activation, which changes slightly with increasing concentration, was defined. A correlation between the behavior of the magnetization of the samples and the temperature dependence of the magnetoresistance and galvanomagnetic properties was found. Using the Hall effect measurements, for X=0.1 a change carrier types by temperature has been found. With increasing concentration to X = 0.3, electrons become dominant.

Acknowledgments
The reported study was funded by RFBR according to the research project №20-52-00005 Bel_a.

References

[1] Petrakovskii G A, Loseva G V, Ryabinkina L I and Aplesnin S S 1995 *JMMM* **140-144** 147
[2] Aplesnin S S, Ryabinkina L I, Romanova O B, Sokolov V V, Pichugin A Y, Galyas A I., Demidenko O F, Makovetski G I and Yanushkevich K I 2009 *Phys. Sol. St.* **51** 698
[3] Aplesnin S S, Romanova O B, Har'kov A M, Balaev D A, Gorev M V, Vorotinov A M, Sokolov V V and Pichugin A Yu 2012 *Phys. St. Sol. (B)* **249** 812
[4] Pal S, Banerjee A, Rozenberg E and Chaudhuri B K 2011 *J. Appl. Phys.* **89** 4955
[5] Aplesnin S S, Ryabinkina L I, Romanova O B, Velikanov D A, Balaev A D, Balaev D A, Yanushkevich K I, Galyas A I, Demidenko O F and Bandurina O N 2008 *J. Exp. Theor. Phys.* **106** 765
[6] Petrakovskii G A, Ryabinkina L I, Velikanov D A, Aplesnin S S, Abramova G M, Kiselev N I and Bobina A F 1999 *Phys. Sol. St.* **41** 1520
[7] Aplesnin S S, Romanova O B, Gorev M V, Velikanov D A, Gamzatov A G and Aliev A M 2013 *J. Phys. Cond. Matt.* **25** 025802
[8] Efrem D'Sa J B C, Bhobe P A, Priolkar K R, Das A, Krishna P S R, Sarode P R and Prabhu R B 2004 *Pramana J. Phys.* **63** 227
[9] Aplesnin S S 1998 *J. Phys. Cond. Mat.* **10** 10061
[10] Aplesnin S S 2000 *Phys. Rev. B* **61** 6780

[11] Aplesnin S S, Romanova O B, Korolev V V, Sitnikov M N and Yanushkevich K I 2017 *J. Appl. Phys.* **121** 7
[12] Aplesnin S S, Romanova O B, Gorev M V, Vasil'ev A D, Demidenko O F, Makovetskii G I and Yanushkevich K I 2012 *Phys. Sol. St.* **54** 1374
[13] Decker D L and Wild R L 1971 *Phys. Rev. B* **4** 3425
[14] Pollard R J, McCann V H and Ward J B 1983 *J. Phys. C: Sol. St. Phys.* **16** 345
[15] Szuszkiewicz B, Hennion B, Witkowska B, Łusakowska E and Mycielski A 2005 *Phys. St. Sol. (C)* **2** 1141
[16] Aplesnin S S, Ryabinkina L I, Romanova O B, Balaev D A, Demidenko O F, Yanuskevich K I and Miroshnichenko N S 2007 *Phys. Sol. St.* **49** 2080
[17] Youn S J, Min B I, Freeman A J 2004 *Phys. Stat. Sol. (B)* **241** 1411
[18] Aplesnin S S, Romanova O B, Sitnikov M N, Kretinin V V, Galyas A I and Yanushkevich K I 2018 *Sem. Sc. and Tech.* **33** 085006
[19] Kriegner D, Výborný K, Olejník K, Reichlová H, Novák V, Marti X, Gazquez J, Saidl V, Nêmec P, Volobuev V V, Springholz G, Holý V and Jungwirth T 2016 *Nature Comm.* **7** 11623
[20] Udod L V, Aplesnin S S, Sitnikov M N and Molokeev M S 2014 *Phys. Sol. St.* **56** 1315
[21] Aplesnin S S, Udod L V, Sitnikov M N and Shestakov N P 2016 *Ceram. Int.* **42** 5177
[22] Aplesnin S S, Ryabinkina L I, Abramova G M, Romanova O B, Kiselev N I and Bovina A F 2004 *Phys. Sol. St.* **46** 2067
[23] Aplesnin S S and Sitnikov M N 2014 *JETP Let.* **100** 95
[24] Aplesnin S S, Romanova O B and Yanushkevich K I 2015 *Phys. St. Sol. (B)* **252** 1792
[25] Aplesnin S S, Bandurina O N, Ryabinkina, L I, Romanova O B, Eremin E V, Gorev M V, Vorotynov A M, Balaev D A, Vasil'ev A D, Galyas A I, Demidenko O F, Makovetskii G I and Yanushkevich K I 2010 *Bull. Russ. Acad. Sci. Phys.* **74** 708
[26] Engle K E, Efrem D'Sa J B C, Das A and Priokar K R 2013 *JMMM* **347** 68

Longitudinal shear waves in an elastic parallelepiped

S Senashov[1], I Savostyanova[1] and O Cherepanova[2]

[1]Reshetnev Siberian State University of Science and Technology, 31 Krasnoyarskii rabochii prospekt, Krasnoyarsk, 660037, Russia
[2]Siberian Federal University, 79 Svobodny Av., Krasnoyarsk, 660041, Russia

E-mail: sen@mail.sibsau.ru

Abstract. This article new solutions will be built which can in particular be used to describe shear waves appearing in the elastic parallelepiped of a finite size. This problem is reduced to three wave equations with two spatial variables. At that, arbitrary initial data will be set, and on the boundary any conditions can be set which belong to either the first, or the second, or the third, or the mixed boundary value problems.

1. Introduction

Many articles and monographs are devoted to the solving of elastic problems. Exceptionally many works are devoted to plane problems, because they reduce to Cauchy-Riemann equations thoroughly studied by mathematicians. Also rather well studied are the problems describing elastic torsion of rods and elastic bending of beams. Less studied are spatial problems, but here are also well-developed methods that use harmonics for solution representation. For these problems there are solutions that describe stress-strain state of space, semi-space limited by a plane etc. If we consider dynamic problems then here the most attention is paid to different types of waves. This is not by accident because it is them that are of the most practical interest. The least studied sphere in the theory of elasticity is the problems describing elastic state of space problems of finite size. This happens because despite of the linearity of the initial equations the solution of boundary value problems for these equations is difficult to carry out [1].

In the suggested paper new solutions are being built which allow us to describe shear waves that can appear in the elastic parallelepiped. This problem is reduced to three wave equations with two spatial variables. At that, arbitrary initial data is set, and on the boundary any conditions can be set which belong to either the first, or the second, or the third, or the mixed boundary value problems.

2. Setting of a problem

Let us consider classical equations of elasticity

$$\begin{aligned}
\rho\frac{\partial^2 u}{\partial t^2} &= (\lambda+G)\frac{\partial e}{\partial x}+G\Delta u,\\
\rho\frac{\partial^2 v}{\partial t^2} &= (\lambda+G)\frac{\partial e}{\partial y}+G\Delta v,\\
\rho\frac{\partial^2 w}{\partial t^2} &= (\lambda+G)\frac{\partial e}{\partial z}+G\Delta w,
\end{aligned} \qquad (1)$$

where u, v, w – displacement vector components, λ, G – elastic constants, $e = \frac{\partial u}{\partial x} + \frac{\partial v}{\partial y} + \frac{\partial w}{\partial z}$ – elastic volume expansion, ρ - density which is considered to be constant.

Equations (1) are well studied. There are many exact solutions built for them and there are many methods for their solving. For System (1) a group of point symmetries is found. This group, by virtue of the linearity of the equations, is infinite-dimensional [2]. On its basis with the use of group analysis technique a series of invariant and functionally invariant solutions is built, which can be used to solve certain physically meaningful problems.

In this work we will be finding solution of Equations (1) in the following form

$$u = u(y, z), v = v(x, z), w = w(x, y) \quad (2)$$

Solutions of such a form were being sought before for the equations of perfect plasticity [3]. This allowed us to obtain new solutions and conservation laws for the equations of plasticity.

3. Finding solutions

Let us insert (2) into System of equations (1). As a result we will obtain the following three wave equations

$$\begin{aligned} \frac{\partial^2 u}{\partial t^2} &= a^2\left(\frac{\partial^2 u}{\partial y^2} + \frac{\partial^2 u}{\partial z^2}\right), \\ \frac{\partial^2 v}{\partial t^2} &= a^2\left(\frac{\partial^2 v}{\partial x^2} + \frac{\partial^2 v}{\partial z^2}\right), \\ \frac{\partial^2 w}{\partial t^2} &= a^2\left(\frac{\partial^2 w}{\partial x^2} + \frac{\partial^2 w}{\partial y^2}\right), \end{aligned} \quad (3)$$

where $a^2 = G / \rho$.

From (2) it follows that $e = \frac{\partial u}{\partial x} + \frac{\partial v}{\partial y} + \frac{\partial w}{\partial z} = 0$, that is why there is no elastic volume expansion and the deformation only consists of shear distortion and rotation [4]. By virtue of this, Equations (3) can be called equations describing shear distortions [4]. From (3) it follows that each equation of this system describes only transverse waves. In this case stress components are equal to

$$\sigma_x = \sigma_y = \sigma_z = 0, \tau_{xy} = G\left(\frac{\partial v}{\partial x} + \frac{\partial u}{\partial y}\right), \tau_{xz} = G\left(\frac{\partial w}{\partial x} + \frac{\partial u}{\partial z}\right), \tau_{yz} = G\left(\frac{\partial w}{\partial y} + \frac{\partial v}{\partial z}\right).$$

In terms of stress components Equations (3) will be written as

$$\begin{aligned} \rho\frac{\partial^2 u}{\partial t^2} &= \frac{\partial \tau_{xy}}{\partial y} + \frac{\partial \tau_{xz}}{\partial z}, \\ \rho\frac{\partial^2 v}{\partial t^2} &= \frac{\partial \tau_{xy}}{\partial x} + \frac{\partial \tau_{yz}}{\partial z}, \\ \rho\frac{\partial^2 w}{\partial t^2} &= \frac{\partial \tau_{xz}}{\partial x} + \frac{\partial \tau_{yz}}{\partial y}. \end{aligned} \quad (4)$$

Note. In stationary case System (4) transforms into a closed system. It contains three equations on three stress components.

$$\frac{\partial\tau_{xy}}{\partial y}+\frac{\partial\tau_{xz}}{\partial z}=0,$$
$$\frac{\partial\tau_{xy}}{\partial x}+\frac{\partial\tau_{yz}}{\partial z}=0,$$
$$\frac{\partial\tau_{xz}}{\partial x}+\frac{\partial\tau_{yz}}{\partial y}=0.$$

Ivlev D.D. called such systems describing the limit state of the body being deformed and he paid special attention to them [5].

4. First boundary value problem for Equations (3)

Let us consider the rectangular parallelepiped $x_0 \le x \le x_1, y_0 \le y \le y_1, z_0 \le z \le z_1, x_1 - x_0 = l_1, y_1 - y_0 = l_2, z_1 - z_0 = l_3$. On the faces of this parallelepiped we will set the first boundary value problem for the first equation of System (3):

$$\begin{aligned}
&u=f_0(y,z),\frac{\partial u}{\partial t}=f_1(y,z),\ for\ t=0,\\
&u=g_1(z,t),\ for\ y=y_0,\\
&u=g_2(z,t),\ for\ y=y_1,\\
&u=g_3(y,t),\ for\ z=z_0,\\
&u=g_4(y,t),\ for\ z=z_1.
\end{aligned}\tag{5}$$

Then its solution is written as [6]

$$\begin{aligned}
u(y,z,t)=&\frac{\partial}{\partial t}\int_{y_0}^{y_1}\int_{z_0}^{z_1} f_0(\xi,\eta)G(y,z,\xi,\eta,t)\,d\xi d\eta+\int_{y_0}^{y_1}\int_{z_0}^{z_1} f_1(\xi,\eta)G(y,z,\xi,\eta)d\xi d\eta+\\
&+a^2\int_0^t\int_{z_0}^{z_1} g_1(\eta,\tau)[\frac{\partial}{\partial\xi}G(y,z,\xi,\eta,t-\tau)]_{\xi=y_0}\,d\eta d\tau-\\
&-a^2\int_0^t\int_{z_0}^{z_1} g_2(\eta,\tau)[\frac{\partial}{\partial\xi}G(y,z,\xi,\eta,t-\tau)]_{\xi=y_1}\,d\eta d\tau+\\
&+a^2\int_0^t\int_{y_0}^{y_1} g_3(\xi,\tau)[\frac{\partial}{\partial\eta}G(y,z,\xi,\eta,t-\tau)]_{\eta=z_0}\,d\xi d\tau-\\
&-a^2\int_0^t\int_{y_0}^{y_1} g_4(\xi,\tau)[\frac{\partial}{\partial\eta}G(y,z,\xi,\eta,t-\tau)]_{\eta=z_1}\,d\xi d\tau,
\end{aligned}$$

where

$$G(y,z,\xi,\eta,t)=\frac{4}{al_1l_2}\sum_{n=1}^{\infty}\sum_{m=1}^{\infty}\frac{1}{\lambda_{mn}}\sin(p_n y)\sin(q_m z)\sin(p_n\xi)\sin(q_m\eta)\sin(a\lambda_{nm}t),$$
$$p_n=\frac{n\pi}{l_1},\ q_m=\frac{m\pi}{l_2},\ \lambda_{nm}=\left(p_n^2+q_m^2\right)^{\frac{1}{2}}.$$

For other equations of System (3) the first boundary value problem is to be set in a similar way, and the solution is to be built as per the above given formula.

The built solution for the first boundary value problem allows us to describe the propagation of transverse elastic waves in the parallelepiped, the faces of which are moving according to the laws defined by Formulas (5) and are initiated by the initial conditions.

Note. Of special interest is the case when all the functions g_i $i = 1, \dots 4$ are equal to zero. This corresponds to the case when the parallelepiped is firmly fixed on all of its faces.

5. Second boundary value problem for Equations (3)

Let us consider the rectangular parallelepiped $x_0 \le x \le x_1, y_0 \le y \le y_1, z_0 \le z \le z_1, x_1 - x_0 = l_1, y_1 - y_0 = l_2, z_1 - z_0 = l_3$. On the faces of this parallelepiped we will set the second boundary value problem for the first equation of System (3)

$$
\begin{aligned}
& u = f_0(y,z), \frac{\partial u}{\partial t} = f_1(y,z), \, for \; t = 0, \\
& \frac{\partial u}{\partial y} = g_1(z,t), \, for \; y = y_0, \\
& \frac{\partial u}{\partial y} = g_2(z,t), \, for \; y = y_1, \\
& \frac{\partial u}{\partial z} = g_3(y,t), \, for \; z = z_0, \\
& \frac{\partial u}{\partial z} = g_4(y,t), \, for \; z = z_1,
\end{aligned}
\tag{6}
$$

where all the functions g_i $i = 1, \dots, 4$ are supposed to be smooth.

Then its solution is written as [6]

$$
\begin{aligned}
u(y,z,t) = & \frac{\partial}{\partial t} \int_{y_0}^{y_1} \int_{z_0}^{z_1} f_0(\xi,\eta) G(y,z,\xi,\eta,t) d\xi d\eta + \int_{y_0}^{y_1} \int_{z_0}^{z_1} f_1(\xi,\eta) G(y,z,\xi,\eta) d\xi d\eta - \\
& -a^2 \int_0^t \int_{z_0}^{z_1} g_1(\eta,\tau) G(y,z,y_0,\eta,t-\tau) d\eta d\tau + \\
& +a^2 \int_0^t \int_{z_0}^{z_1} g_2(\eta,\tau) G(y,z,y_2,\eta,t-\tau) d\eta d\tau - \\
& -a^2 \int_0^t \int_{y_0}^{y_1} g_3(\xi,\tau) G(y,z,\xi,z_0,t-\tau) d\xi d\tau + \\
& +a^2 \int_0^t \int_{y_0}^{y_1} g_4(\xi,\tau) G(y,z,\xi,z_1,t-\tau) d\xi d\tau,
\end{aligned}
\tag{7}
$$

where

$$
G(y,z,\xi,\eta,t) = \frac{t}{l_1 l_2} + \frac{2}{a l_1 l_2} \sum_{n=1}^{\infty} \sum_{m=1}^{\infty} \frac{A_{nm}}{\lambda_{mn}} \cos(p_n y) \cos(q_m z) \cos(p_n \xi) \cos(q_m \eta) \sin(a \lambda_{nm} t),
$$

$$p_n = \frac{n\pi}{l_1}, q_m = \frac{m\pi}{l_2}, \lambda_{nm} = \left(p_n^2 + q_m^2\right)^{\frac{1}{2}},$$

$$A_{nm} = \begin{cases} 0, for \quad n = m = 0, \\ 1, for \; nm = 0, (n \neq m), \\ 2, for \quad nm \neq 0. \end{cases}$$

For other equations of System (3) the second boundary value problem is to be set in a similar way, and the solution is to be built as per the above given formula.

The second boundary value problem for Equation (3) means that on the faces of the parallelepiped the shifts in the direction of the corresponding derivatives are defined.

6. Third boundary value problem for Equations (3)

Let us consider the rectangular parallelepiped $x_0 \leq x \leq x_1, y_0 \leq y \leq y_1, z_0 \leq z \leq z_1, x_1 - x_0 = l_1, y_1 - y_0 = l_2, z_1 - z_0 = l_3$. On the faces of this parallelepiped we will set the third boundary value problem for the first equation of System (3):

$$\begin{aligned} & u = f_0(y,z), \frac{\partial u}{\partial t} = f_1(y,z), for \; t = 0, \\ & \frac{\partial u}{\partial y} + k_1 u = g_1(z,t), for \; y = y_0, \\ & \frac{\partial u}{\partial y} + k_2 u = g_2(z,t), for \; y = y_1, \\ & \frac{\partial u}{\partial z} + k_3 u = g_3(y,t), for \; z = z_0, \\ & \frac{\partial u}{\partial z} + k_4 u = g_4(y,t), for \; z = z_1, \end{aligned} \tag{8}$$

where k_i $i = 1, \ldots, 4$ some constants.

Then its solution is written as (7), where

$$G(y,z,\xi,\eta,t) = \frac{4}{a}\sum_{n=1}^{\infty}\sum_{m=1}^{\infty}\frac{1}{E_{nm}\left(\mu_n^2 + \nu_m^2\right)^{\frac{1}{2}}}\sin(\mu_n y + \varepsilon_n)\sin(\nu_m z + \sigma_m) \times$$

$$\times \sin(\mu_n \xi + \varepsilon_n)\sin(\nu_m \eta + \sigma_m)\sin(at\left(\mu_n^2 + \nu_m^2\right)^{\frac{1}{2}}),$$

$$\varepsilon_n = arctg\,\frac{\mu_n}{l_1},\; \sigma_m = arctg\,\frac{\nu_m}{l_2},\; E_{nm} = [l_1 + \frac{(k_1k_2 + \mu_n^2)(k_1 + k_2)}{(k_1^2 + \mu_n^2)(k_2^2 + \mu_n^2)}]\,[l_2 + \frac{(k_3k_4 + \nu_m^2)(k_3 + k_4)}{(k_2^2 + \nu_n^2)(k_4^2 + \nu_m^2)}]$$

where μ_n, ν_n – positive roots of transcendental equations

$$\mu^2 - k_1k_2 = (k_1 + k_2)\mu ctg\,(l_1\mu),\; \nu^2 - k_3k_4 = (k_3 + k_4)\nu ctg\,(l_2\nu).$$

7. Conclusion

In this work presented is a class of new solutions of classical equations of elasticity. These solutions can be used not only to describe the development of shear waves in elastic bodies but also to describe stress-strain state of an elastic parallelepiped being twisted around three orthogonal axes as it is fulfilled in [3] in the case of plastic deformation.

References

[1] 1984 *Encyclopaedia of Mathematics* (Moscow: Soviet Encyclopaedia) **5** p 1246

[2] Annin B D, Bytev V O and Senashov S I 1985 *Group Properties of Equations of Elasticity and Plasticity* (Novosibirsk: Nauka) p 140

[3] Senashov S I and Filushina E V 2017 On limit state of bodies being deformed *Informational technologies and Mathematical modelling* (Krasnoyarsk) pp 73-93.

[4] Timoshenko S P and Goodyear J 1975 *Theory of Elasticity* (Moscow: Nauka) p 570

[5] Ivlev D D, Maximova L A and Nepershin R I 2008 *Limit state of deformed bodies and subsurface rocks* (Moscow:Phizmatlit) p 832

[6] Polyanin A D 2001 *Reference book on linear equations of mathematical physics* (Moscow: Phizmatlit) p 576

IOP Conf. Series: Materials Science and Engineering 1230 (2022) 012018 doi:10.1088/1757-899X/1230/1/012018

Applications of group theory in crystallography

V I Senashov

Siberian Federal University, 79 Svobodny Av., 660041 Krasnoyarsk, Russia
Institute of Computational Modelling SB RAS, 50/44 Akademgorodok St., 660036 Krasnoyarsk, Russia

E-mail: sen1112home@mail.ru

Abstract. Group theory is a powerful tool for studying symmetric physical systems. Such systems include, in particular, molecules and crystals with symmetry. Group theory serves to explain the most important characteristics of atomic spectra. Group theory is also applied to the problems of atomic and nuclear physics. This paper gives examples of the use of the apparatus of group theory in research on crystallography, quantum mechanics, elementary particle physics. In particular, in these studies matrix groups and representations of unitary groups are actively used. For such groups we give an overview of the results on their recognition by the spectrum (by the orders of the elements of the group). This direction has been intensively developed in recent years both in our country and abroad. Recognition of finite simple non-Abelian groups by spectrum has been studied for last thirty years in Yekaterinburg at the Institute of Mathematics and Mechanics of the Ural Division of the Russian Academy of Sciences, in Chelyabinsk Federal University and in the Novosibirsk Institute of Mathematics of Siberian Division of the Russian Academy of Sciences. Some simple non-Abelian groups are not recognizable by their spectra. We have proposed an approach for recognizing groups by the bottom layer. The bottom layer of a group is the set of its elements of prime orders. A group is called recognizable by the bottom layer under additional conditions if it is uniquely restored by the bottom layer under these conditions. The paper considers some examples of simple non-Abelian finite groups that are not recognizable by spectra. For these examples, simultaneous recognition by spectrum and by the bottom layer is proved.

1. Examples of the application of group theory in crystallography and related fields

Many publications have been devoted to applications of group theory to crystallography, quantum mechanics, atomic and nuclear physics, elementary particle physics (see, for example, [1–6]).

Group theory is a powerful tool for studying symmetric physical systems. Such systems include, in particular, molecules and crystals. The monograph [1] examines the application of group theory to the study of the vibrations of atoms that make up a molecule relative to their equilibrium positions.

In 1867, a Russian scientist, academician and professor of the Petersburg Artillery School A.V. Gadolin (1828–1892) gave a clear description of 32 crystallographic groups [2].

In quantum mechanics, the state of a system is completely determined by the wave function or operator. If the set of eigenvalues (operator spectrum) is discrete, we are dealing with the quantization phenomenon when the measured values of the operator are discrete.

The symmetry group G is usually determined by several elements (one or two), raising the latter to a power or multiplying them until new symmetry elements appear. Moreover, it may turn out that the group constructed in this way is only a subgroup of the complete group G. To avoid this, it is necessary to use all available information about the symmetry of the system.

In most cases, rotations, reflections and translations act as elements of G. In the presence of system symmetry, group theory provides a convenient method for classifying its eigenfunctions and eigenvalues. If the system has symmetry, then under the action of the group's symmetry operations its eigenfunctions are transformed according to irreducible representations.

A physical system that exists due to the attraction between its components has a ground state with maximum symmetry. Mathematically, this means that the ground state must be invariant with respect to all operations of the group.

Crystals, like molecules, have symmetry. This means that the atoms or ions that make up the crystal are arranged in an orderly spatial lattice.

In the general case, a certain set of atoms with a specific arrangement is associated with each lattice site. Although real crystals are finite, they are nevertheless limited to considering the model of an infinite crystal. The symmetry of an infinite crystal is determined, as already mentioned, by both rotations and translations combining the crystal with itself and forming the so-called crystalline groups.

E. Wigner [3] notes that group theory serves to explain the most important characteristics of atomic spectra. When considering specific problems, group theory allows us to draw conclusions about the behavior of the system without complex calculations using ideas about the symmetry of the system. Such predictions are essential in the study of spectra. If the interaction of electrons is considered as a perturbation, then the spectrum levels are split. Regarding the resulting levels, their symmetry properties are known. They manifest themselves in the transformational properties of the corresponding eigenfunctions with respect to the rearrangement of electrons, rotations, and reflections. Therefore, each level corresponds to three representations: one representation of a symmetric group, one is a group of rotations, and one is a group of reflections.

Most importantly, optical transitions occur only between levels that have certain properties or rules. These rules should follow from quantum mechanics. In the study of atomic spectra, symmetric groups and unitary matrix groups are used. In this case, non-Abelian groups of rotations and reflections are studied. Various representations of unitary groups are also being studied.

Group theory has now found application in many fields of physics [4]. In particular, group theory has applications in elementary particle physics. The main role in modern particle physics are played the unitary groups. These groups are the groups of spin and isotopic transformations, and also forms the basis of the transformation groups of weak interactions. The unitary group $SU(3)$ is the basis of the unitary symmetry model.

Thus, the issue of particle classification is closely related to the representations of unitary groups.

Other important examples of the specific application of group theory and their representations are examples of calculations of such important characteristics of elementary particles as magnetic moments and axial-vector coupling constants in a unitary symmetry model [5].

Group theory is also applied to the problems of atomic and nuclear physics. One of the main tasks of atomic physics and nuclear physics is to determine the energy levels of a system of equivalent particles. Since it is impossible to precisely solve this problem for a system of interacting particles, one has to resort to perturbation theory methods. In this case, the perturbation will consist of some part of the field of one particle plus the interaction between the particles. If the particles are identical, then the interaction operator will be symmetric over all particles.

Another point of view is as follows. The wave function is considered as a vector in n-dimensional space spanned by basis vectors. If the basis vectors are subjected to a unitary transformation, we get another basis for the same vector space. In addition, unitary transformations can be made unimodular. As a result, we can assume that the space spanned by functions gives us a basis for representing the unitary unimodular group $SU(n)$ [6].

2. The study of simple non-Abelian groups using their spectra

Recognition of finite simple non-Abelian groups by their spectra has been studied for last thirty years in Yekaterinburg at the Institute of Mathematics and Mechanics, in Ural Branch of the Russian

Academy of Sciences, Chelyabinsk Federal University, and in the Novosibirsk Institute of Mathematics of Siberian Division of the Russian Academy of Sciences: A. S. Kondratiev, V. D. Mazurov, A. V. Vasiliev, A. A. Buturlakin, A. M. Staroletov, M. R. Aleeva, O. A. Alekseeva, I. B. Gorshkov, M. A. Grechkoseeva, A. V. Zavarnitsin and others. Also, this topic was developed abroad by R. Brantl, W.J. Shi, G. Y. Chen, N. Chigira, H. W. Deng, N. D. Gupta, C. E. Praeger, M. R. Darafsheh, A. R. Moghaddamfar [7–22].

Definition. The *spectrum* $\omega(G)$ of a finite group G is called the set of all element orders of G.

In other words, a natural number n is in $\omega(G)$ if and only if there is an element of order n in G.

Definition. A finite group G is called *recognizable by its spectrum* if any finite group whose spectrum coincides with the spectrum of G is isomorphic to G [16].

Among the results on recognizability of simple non-Abelian groups, a typical result is the following theorem of A.V. Vasiliev:

Let G be a finite simple group $L_6(3)$ and H be a finite group with the property $\omega(H) = \omega(G)$. Then H is isomorphic to G or H is isomorphic to $G(\gamma)$, where γ is a graph automorphism of the group G of order 2. In particular, $h(\mathrm{G}) = 2$ [9].

We say that for a finite group G the recognition problem is solved if we know the value of $h(\omega(G))$ (for brevity, $h(G)$).

More precisely, G is said to be *recognizable by its spectrum* in such designations if $h(G) = 1$, *almost recognizable by spectrum* if $1 < h(G) < \infty$, and *nonrecognizable by spectrum* if $h(G) = \infty$.

Almost all simple non-Abelian groups are recognizable by their spectra. But there are exceptions: different groups of this set have the same spectra.

3. The study of simple non-Abelian groups using the bottom layer

Recognition of groups by bottom layer has been studied in Krasnoyarsk for last five years by author.

We will show the possibility of recognizing by the bottom layer for such simple non-Abelian groups with the same spectrum on the example of the groups $S_6(2)$ and $O_{8^+}(2)$.

Groups $S_6(2)$ and $O_{8^+}(2)$ are unrecognizable by their spectra [9].

Definition. The *bottom layer* of a group is the set of its elements of prime orders.

Previously, recognizability by the bottom layer of groups in the class of infinite groups with additional finiteness conditions was considered by the author together with I. A. Parashchuk. As additional restrictions to the task by the bottom layer of the group, the completeness of the group and its layer-finiteness were used [23].

Recall that a group is called *layer-finite* if it has a finite number of elements of each order [24].

In this paper, instead of such finiteness conditions, we consider the simplicity of a group, non-Abelianity and its finiteness.

Definition. A group is called *recognizable by the bottom layer* under additional conditions if it is uniquely restored by the bottom layer under these conditions. A group G is said to be *almost recognizable by the bottom layer* if there are a finite number of pairwise non-isomorphic groups with the same bottom layer as in the group G. A group G is said to be *unrecognizable by the bottom layer* if there exist an infinite number of pairwise non-isomorphic groups with the same bottom layer such as in the group G [23].

By the bottom layer of a group, you can sometimes recognize a group, sometimes you can say something about the properties of such a group. Among the results of groups recognized by the bottom layer with the additional condition, that the groups without a unit element coincide by the bottom layer are the following:

– if the bottom layer of the group consists of elements of order 2 and the group does not have non-identity elements of other orders, then G is an elementary Abelian 2-group;

– if the bottom layer consists of elements of orders 2, 3, 5 and the group has no nonidentity elements of other orders, then A. S. Kondratiev and V. D. Mazurov proved that this is a group of even permutations on five elements [21].

Many results for groups with a given bottom layer describe only some of the group properties. For example, V. D. Mazurov proved that a group with a bottom layer consisting of elements of orders 2, 3, 5, in which all other non-identity elements are of order 4 with the additional condition that the groups coincide without an identity element with the bottom layer, is locally finite [22].

The almost recognizable by bottom layer groups in the class of infinite layer-finite groups are the groups in the following example. V. P. Shunkov proved that if the bottom layer in an infinite layer-finite group consists of one element of order 2, then the group G is either a quasicyclic or an infinite generalized group of quaternions [25].

We will show that the groups $S_6(2)$ and $O_{8^+}(2)$ are recognizable simultaneously by the spectrum and by the bottom layer.

The group $O_{8^+}(2)$ is simple, it has order $174182400 = 2^{12} \cdot 3^5 \cdot 5^2 \cdot 7$. Using the GAP application package, it was established that it has 69615 involutions (elements of order 2) and 24883200 elements of order 7.

The group $S_6(2)$ is simple, it has order $1451520 = 2^9 \cdot 3^4 \cdot 5 \cdot 7$. Using the GAP application package, it was found that it has 5103 involutions (elements of order 2) and 207360 elements of order 7.

That is, these groups differ in the bottom layer, but do not differ in the set of orders of elements.

As established by A. V. Vasiliev [9], a finite simple group U_4 (5) is not recognizable by its spectrum. In this regard, we prove that simultaneous recognition of this group by spectrum and bottom layer is possible.

Let G be the finite simple group U_4 (5) and H a finite group with the property $\omega(H) = \omega(G)$ and the bottom layer is the same as that of the group U_4 (5). Then H is isomorphic to G. That is, the group U_4 (5) is unique with such a spectrum and a bottom layer.

Indeed, let G be the finite simple group U_4 (5) and H be a finite group with the property $\omega(H) = \omega(G)$. By the theorem of A. V. Vasiliev [9], in addition to the group U_4 (5), there is only one such group H isomorphic to $G(\gamma)$, where γ is a field automorphism of the group G of order 2. The groups $U_4(5)$ and H have the same spectrum, at the same time these groups have different bottom layers, which differ at least by an element of order 2.

Thus, we have established that the group $U_4(5)$ is unique with such a spectrum and a bottom layer.

4. Conclusion

The paper gives some examples of applications of finite groups in crystallography, quantum mechanics, atomic and nuclear physics, elementary particle physic. A small review of the results on the recognition of simple non-Abelian groups by their spectra is presented. Previously, recognizability of the bottom layer groups in the class of infinite groups with additional finiteness conditions was considered. In this paper, instead of such finiteness conditions, we consider the simplicity of a group, non-Abelianity and its finiteness. The recognizability of some simple non-Abelian groups by both their spectra and their bottom layer for groups that are not recognizable by their spectra is established. In particular, we proved that the groups $S_6(2)$ and $O_{8^+}(2)$ which can not be recognized by spectrum alone can be recognized by the combination of their of their spectrum and bottom layer. The same is proved for the group $U_4(5)$.

These results will find applications in the study of finite groups, in particular, matrix groups. Matrix groups, in turn, find application in crystallography, atomic and nuclear physics and elementary particle physics.

References

[1] Ivanov E N 2006 *The Theory of Groups and its Applications in Physics* (Moscow: MIET) p 160
[2] Kartonova LV 2015 *Fundamentals of Crystallography* (Vladimir: Vladimir State University named after A.G. and N.G. Stoletovs) p 80
[3] Wigner E 1961 *Group Theory and its Applications to the Quantum Mechanical Theory of Atomic Spectra* (Moscow: Inostrannaya Literatura)
[4] Lyubarsky G Ya 1957 *Group Theory and its Application in Physics* (Moscow: Gostekhizdat)

[5] Artamonov V A and Slovokhotov Yu L 2005 *Groups and Their Applications in Physics, Chemistry, Crystallography: Textbook for students of higher institutions* (Moscow: Publishing Center «Academy») p 512
[6] Hammermesh M 1966 *Group Theory and its Application to Physical Problems* (Moscow: Mir) p 588
[7] Zavarnitsin A V and Mazurov V D 1999 *Algebra and Logic* **38(3)** 296–315
[8] Buturlakin A A and Vasiliev A V 2014 *Algebra and Logic* **53(4)** 541–4
[9] Vasiliev A V 2005 *Sib. matem. journal* **46(2)** 315-24
[10] Vasiliev A V and Staroletov A M 2014 *Algebra and Logic* **53(6)** 669–92
[11] Gorshkov I B 2014 *Algebra and Logic* **53(6)** 693–703
[12] Shi W 1987 *Adv. Math.* **16** 397–401
[13] Brandl R and Shi W1994 *J. Algebra* **163(1)** 109–14
[14] Brandl R and Shi W 1991 *J. Algebra* **143(2)** 388–400
[15] Praeger C E and Shi W 1994 *Commun. Algebra* **22(5)** 1507–30
[16] Mazurov V D 1998 *Algebra and Logic* **37(6)** 651–66
[17] Mazurov V D 2005 *Izv. of Ural. State University* **36(Matem. Mekhan., Issue 7)** 119-138
[18] Mazurov V D 1997 *Algebra and Logic* **36(1)** 37–53
[19] Darafsheh M R and Moghaddamfar A R 2000 *Algebra Colloq.***7(4)** 467–76
[20] Zavarnitsin A V 2002 *Sibirsk. matem. journal* **43(5)** 1002–6
[21] Kondratiev A S and Mazurov V D 2000 *Sib. matem. journal* **41(2)** 360–71
[22] Mazurov V D 2000 *Algebra and Logic* **39(3)** 189–98
[23] Parashchuk I A and Senashov V I 2018 *Siberian Mathematical Journal of Science and Technology* **19(2)** 223–6
[24] Senashov V I 1993 *Layer-finite groups* (Novosibirsk: Science, Siberian publishing group) p 159
[25] Shunkov V P 1970 *Algebra and Logic* **9(4)** 484–96

The thermodynamic analysis of physico-chemical processes of microarc oxidation of aluminium and titanium alloys

A A Snezhko, T V Strekaleva, I V Trifanov, T A Pchelkina and E A Karelina

Reshetnev Siberian State University of Science and Technology, 31 Krasnoyarskii rabochii prospekt, Krasnoyarsk, 660037, Russia

E-mail: sibgau-uks@mail.ru

Abstract. The paper presents the results of assessing the probability of the occurrenceof physicochemical reactions in the electrolyte and on the electrodes while forming oxide coatings on aluminium and titanium alloys usinga microarc method. Calculations have been performed with the help of the thermodynamic analysis using the Hess's law and Ostwald step rule. The calculations are coherent with the research results.

1. Introduction

To develop the techniques of protective coating of machine component parts using microarc oxidation, it is necessary to study possible versions of chemical reactions ensuring the predicted composition and technological properties of coatings.

Microarc oxidation (MAO) is a relatively new type of surface treatment and hardening, mainly of metallic materials. It derives from traditional anodizing and relates to electrochemical processes. A distinctive feature of microarc oxidation is the participation of surface microdischarges in the process of coating formation.

The coating made by microarc oxidation resembles ceramics in its properties. Such a surface layer has high hardness and wear resistance. It scarcely interacts with corrosion environments. That allows us to use it as multifunctional protection of valve metals (aluminium, titanium, tantalum, vanadium, tungsten, niobium, etc.) in aircraft construction, rocket building, as well as in automobile, textile and other industries.Valve metals are the metals whose oxides form such films that their conductivity in opposite directions varies greatly [1].

It is possible to model the composition of a coating and its properties theoretically through thermodynamic parameters and / or practically through technological ones. Most authors are guided by the optimal technological parameters selected in almost every individual case. Thatretains the implementation of microarc oxidation in enterprises with the limited set of initial and boundary conditions [2].

2. Methods and materials

Coatings were formed by the MAO method on aluminum alloys D16T, Amg6, AK9ch with a cathode-anode ratio of 1 and 1.3 and current densities of 10–30 A / dm^2 for 90 min. Samples with dimensions of 20 x 20 x 5 mm were used as the substrate. The MAO of the samples was carried out in a slightly alkaline aqueous electrolyte containing potassium hydroxide KOH with a concentration of 1 g / L and liquid glass Na_2SiO_3 (10 g / L). The mass gain of the sample after MAO was measured on an analytical balance. Coating thicknesses (total and solid layer) were monitored on transverse sections using an MBI-15 optical microscope, the microstructure of the coatings was studied using a Jenavert optical microscope with magnification up to 1000 and a SEM 200 electron microscope. The oxide layer was etched in boiling phosphoric acid. The microhardness of the coatings was determined by the standard method on a PMT-3 microhardness meter with an indenter load of 0.981 N (100 g). The qualitative phase composition of the coatings was studied on a DRON-3.0 X-ray diffractometer.

3. Results and discussion

The set of various galvanic processes that occur at high temperatures and the transport of a matter in an electric arc created between electrodes placed in a medium with ionic or electron-ionic conductivity is characterized by the extremely nonequilibrium condition of microarc oxidation.

It is quite problematic to evaluate the phase composition of the coatings by analyzing the possible directions of chemical reactions for nonequilibrium conditions due to the multifactorityof the process and its parallel occurrence in several aggregate states (solid, liquid, gaseous, and plasma). Therefore, for a rough estimate, we can limit ourselves to equilibrium conditions.

To determine the most probable chemical reactions that predict the composition of a coating, it makes sense to apply the criterion of the spontaneous occurrence of a chemical reaction in an non-insulated physicochemical system for isobaric conditions – the change in the Gibbs energy of the $\boldsymbol{\Delta G_{react}}$reaction. The change in the Gibbs energy of a thermodynamic system is presented by the following formula:

$$\Delta G = \Delta H - T \cdot \Delta S, \quad (1)$$

where ΔH is the change of enthalpy, ΔS is the change of entropy, T is absolute temperature.

In the course of the work, the possible reaction versions for aluminium and titanium alloys have been analyzed by determining the $\Delta G°_{react}$ reaction. The reaction is considered possible with the negative value of ΔG [3, 4]. The results are presented in the form of the graphs for titanium and aluminium alloys in figures 1-3, 4-5, respectively.

The following versions of reactions are possible:

In the electrolyte:

$$Na_2SiO_3 \leftrightarrow 2Na^+ + SiO_3^{2-}$$
$$2SiO_3^{2-} \rightarrow 2SiO_2^{2-} + O_2$$
$$H_2O \leftrightarrow H^+ + OH^-$$
$$KOH \leftrightarrow K^+ + OH^-;$$

At the cathode:

$$6H^+ + 6e \rightarrow 3H_2\uparrow;$$

At the anode:

1) version:

$$2Al \rightarrow 2Al^{3+} + 6e$$
$$Al^{3+} + 3OH^- \rightarrow Al(OH)_3$$
$$Al(OH)_3 + OH^- \rightarrow AlO_2^- + 2H_2O$$
$$AlO_2^- + K^+ \rightarrow KAlO_2$$
$$2Al + 2KOH + 2H_2O \rightarrow 2KAlO_2 + 3H_2\uparrow$$
$$2KAlO_2 + 4H_2O \leftrightarrow \mathbf{Al_2O_3 \cdot 3H_2O} + 2KOH;$$

2) version:

$$2Al \rightarrow 2Al^{3+} + 6e$$
$$Al^{3+} + 3OH^- \rightarrow Al(OH)_3$$
$$2Al(OH)_3 \rightarrow \mathbf{Al_2O_3 \cdot 3H_2O};$$

3) version:

$$4Al + 3O_2 \rightarrow \mathbf{2Al_2O_3}$$

To obtain coatings by microarc oxidation with increased hardness, during chemical reactions it is necessary to reach temperatures that make it possible to obtain α-Al_2O_3, which is possible at high temperatures,approximately in the range 1400–1500 °C with the final degeneration of the γ modification [5]. Obviously, the first group of reactions is a forbiddenversion in this direction because of the failure of the final reaction of the formation of alumina (figures 1). There are some limitations on the second version of the formation of alumina as well. Reactions are allowed, only in the range 0 ... 700 ° C (figures 2). The formation of the coatings with α-Al_2O_3 is completely allowed in the third version of the formation of alumina (figures 3).

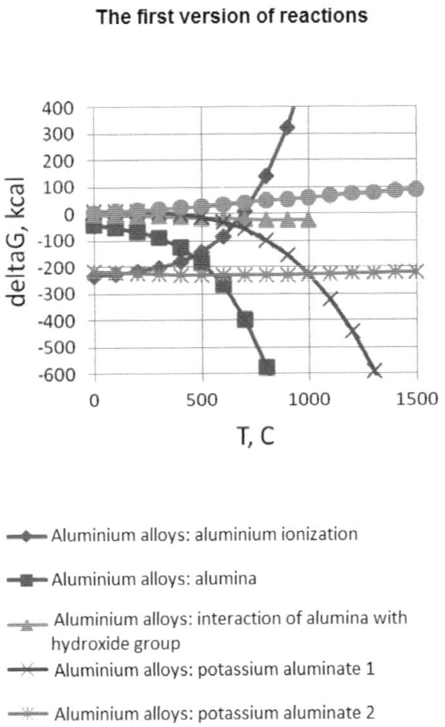

Figure1. The dependence of the change of Gibbs energy on the temperature during the reactions while forming coatings by microarc oxidation on aluminium alloys (first version of reactions).

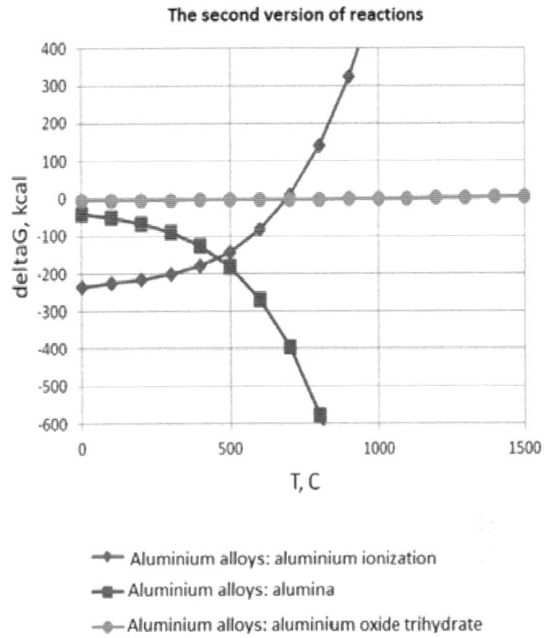

Figure2. The dependence of the change of Gibbs energy on the temperature during the reactions while forming coatings by microarc oxidation on aluminium alloys (second version of reactions).

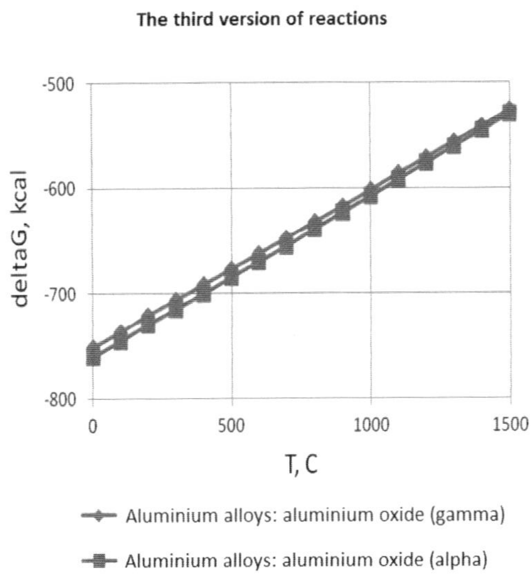

Figure3. The dependence of the change of Gibbs energy on the temperature during the reactions while forming coatings by microarc oxidation on aluminium alloys (third version of reactions).

IOP Conf. Series: Materials Science and Engineering 1230 (2022) 012019 doi:10.1088/1757-899X/1230/1/012019

The negative $\triangle G^{\circ}$ value while forming coatings by microarc oxidation on titanium alloys according to the first version is characteristic for all the reactions except the formation of titanium dioxide (thermodynamically forbidden). As for the second method, we have found the high probability of reactions in the right direction (figures 4and 5). Kinetically, in accordance with the Ostwald step rule, the most probable is the reaction proceeding with the smallest in modulus$\triangle G^{\circ}$ value. One of these reactions is the anatase formation reaction, which occurs first of all. Further, the part of the anatase is modified into rutile, a harder phase [6].

In the electrolyte:

$$Na_2SiO_3 \leftrightarrow 2Na^{+} + SiO_3^{2-}$$
$$2SiO_3^{2-} \rightarrow 2SiO_2^{2-} + O_2$$
$$H_2O \leftrightarrow H^{+} + OH^{-}$$
$$KOH \leftrightarrow K^{+} + OH^{-};$$

At the cathode:

$$4H^{+} + 4e \rightarrow 2H_2\uparrow;$$

At the anode:

1) version:

$$Ti \rightarrow Ti^{4+} + 4e$$
$$Ti^{4+} + 4(OH)^{-} \rightarrow Ti(OH)_4$$
$$Ti + 2KOH + H_2O \rightarrow K_2TiO_3 + 2H_2\uparrow$$
$$K_2TiO_3 + 2H_2O \leftrightarrow \mathbf{TiO_2 \times H_2O} + 2KOH;$$

2) version:

$$Ti + O_2 \rightarrow \mathbf{TiO_2};$$

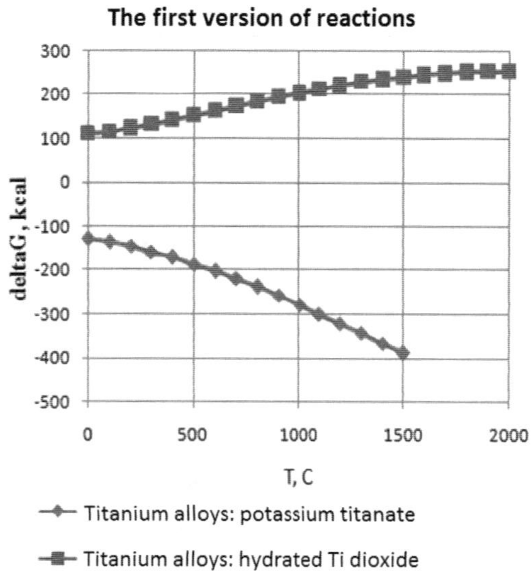

Figure 4. The dependence of the change of Gibbs energy on the temperature during the reactions while forming coatings by microarc oxidation on titanium alloys (first version of reactions).

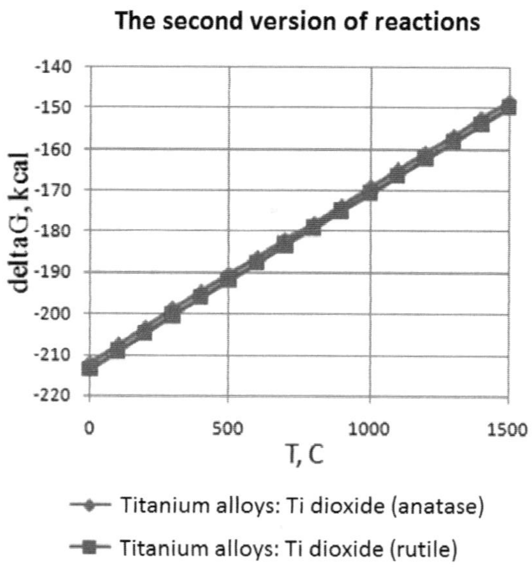

Figure 5. The dependence of the change of Gibbs energy on the temperature during the reactions while forming coatings by microarc oxidation on titanium alloys (second version of reactions).

According to the Ostwald step rule and Gibb's test,the reaction of forming aluminaof γ-phase is more energetically beneficial. During the heat treatment of aluminium hydroxides at the temperature 400 ° C, a cubic γ-shape is obtained. The irreversible conversion to α-Al_2O_3 occurs at 1100–1200 ° C with γ-modification, however, the rate of this process is slow, and to complete the phase transition, either mineralizers or the increase of treatment temperature to 1400–1450° C are required [5].

4. Conclusion

We have carried out the thermodynamic analysis of physicochemical processes of microarc oxidation. The analysis shows thatthe formation of coatings for aluminium alloysoccursinthe third version;for titanium alloys - in the second version. It is offered to form the coatings received by microarc oxidation with a preset structure and physicomechanical propertiesby regulatingcurrent density within the scope of 25-27,5 amperes per square decimeter [2], which correlates to the chemical reaction temperature in the breakdown pathand ensures hard phases in the composition of a coating (for aluminium-α-Al_2O_3, for titanium – rutile).

References

[1] Valve metals *Kommersant*, available at: https://www.kommersant.ru/doc/3780225
[2] Golenkova A A 2006 *Improving the technology of forming wear-resistant coatings on aluminium alloys by microarc oxidation* (Krasnoyarsk: Reshetnev University) p 15
[3] Volmer M 1986 The kinetics of a new phase formation (Moscow: FIZMATLIT) p 208
[4] Zhereb V and Golenkova A 2007 *Thermodynamic analysis of physicochemical systems* (Krasnoyarsk: Reshetnev University) p 34
[5] Paglia G 2004 *Determination of the Structure of γ-Alumina using Empirical and First Principles Calculations Combined with Supporting Experiments* (Perth, Curtin University of Technology)
[6] 1998 *Chemical Encyclopedia. TRI-YATR* vol 5, ed N S Zefirov et al (Moscow: Big Russian Encyclopedia) p 783

IOP Conf. Series: Materials Science and Engineering 1230 (2022) 012020 doi:10.1088/1757-899X/1230/1/012020

An intelligent voice recognition system based on fuzzy logic and the bag-of-words technique

O S Stefanenko, L V Lipinskiy, A S Polyakova, J A Khudonogova and E S Semenkin

Reshetnev Siberian State University of Science and Technology, 31 Krasnoyarskii rabochii prospekt, Krasnoyarsk, 660037, Russia

E-mail: polyakova-o.s@mail.ru, lipinskiyl@mail.ru , polyakova_nasty@mail.ru, j.khudonogova@gmail.com, eugenesemenkin@yandex.ru

Abstract. This paper describes a method for recognizing voice command based on a fuzzy logic system capable of perceiving fuzzy commands, i.e. commands containing fuzzy terms, for example, ‘close’, ‘closer’, ‘close to’, ‘closer than’, ‘further’ and ‘very far’. The developed approach has the ability to be trained for a specific user. The developed fuzzy logic system is used to recognize linguistically inaccurate commands in order to increase the expressiveness of the language for control of a moving robot.

1. Introduction

Robotics is no longer an elitist field of science, and the usage of robots is no longer limited to highly specialized tasks [1]. However, one of the constraints to the development of robotics is the limited capabilities of human-machine communication [2]. Classic computer manipulators, such as a keyboard, mouse, or sensors, are ineffective when working with robots.

In recent times, the number of studies devoted to human-machine interaction has been increasing rapidly [3]. This field of knowledge contains many disciplines in the field of control, machine learning and others, but does not describe one of it. At the same time, the task of voice control is one of the most critical in general human-machine interaction, since the state of the robot must change as a result of the voice command. It is not always possible to switch to an operator, such as in automated information systems. In addition, voice command often need to be recognized and interpreted in real time, which imposes certain restrictions on the voice control system.

When discussing voice control technologies, it is necessary to distinguish between speech recognition systems [4] and voice-based control systems [5], which, unlike the former, not only directly recognize speech, but also interpret the command and translate it into machine signals. At the same time, commands are rarely expressed in single words, and are more often represented as whole phrases. Changing the word order in these phrases can partially or completely change the meaning.

Classification algorithms for solving the voice command recognition problem can be divided into two large groups: statistical and probabilistic.

The most well-known statistical approach to speech recognition, which is well-established in voice identification and voice control tasks, is the hidden Markov model (HMM). HMM is based on the concept of Markov chains, which can represent any random sequential process that undergoes transitions from one state to another [6].

HMMs allow to perform preliminary formation of a reference database of words or phrases and then to identify the presented sound images based on comparison with this database [6, 7]. One of the types of HMMs is continuous HMMs, where transitions between hidden states and the arrival of observations can occur at arbitrary points of time [8, 9].

Artificial neural networks, starting from feed forward and back-propagation neural networks, multi-layer perceptron (MLP), Hopfield networks (HN), and other more complex variants were used as probabilistic classifiers [10, 11, 12, 13]. Convolutional neural networks (CNN) and deep convolutional neural networks (DCNN) are very different from other types of networks, which are usually used for image processing, less often for audio, but they show an improvement compared to classic MLPs [14].

The complexity is in the use of evaluation words with linguistic uncertainty, such as 'faster', 'slower', 'stronger' and 'louder', which greatly complicates the interpretation of the command and the generation of machine control signals.

In addition, for command recognition it is important not only to recognize speech, but also to interpret the received words as a command (taking into account linguistically inaccurate expressions). The development of such an interpreter requires a separate study in order to develop an effective interpretation model and develop a procedure for setting numerical parameters.

In view of the foregoing, in this paper we propose the development of algorithmic support to increase the expressiveness of language commands in voice control systems by introducing decision support systems in the fuzzy logic system for the operation of robots moving in space.

2. Process description

Formally, the task of voice control of a given object can be represented as a purposeful change in the state of an object using voice command (figure 1).

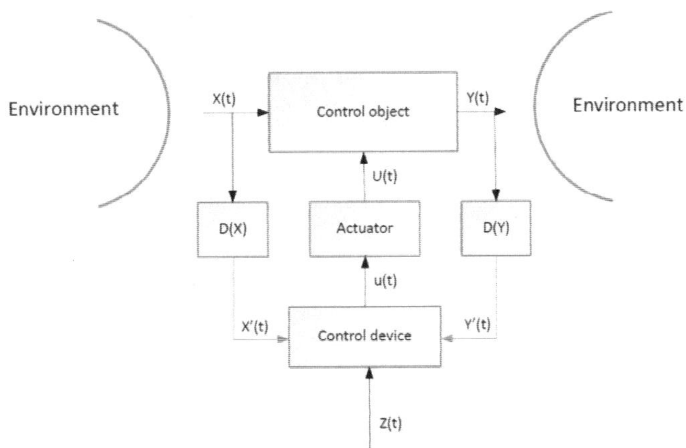

Figure 1. General control scheme.

Figure 1 shows the classic control scheme of a complex object. The control object is that part of the world around us the state of which we want to control. This object has inputs *X(t)*, outputs *Y(t)* and some control channels *U(t)*. A complex object is controlled using a control device. As a rule, this is an electronic module based on a microcontroller. This module accepts the control target *Z(t)* as the main input. Using sensors *D(X)* and *D(Y)*, it collects information about the state of the control object and generates a control action in the form of some machine command *u(t)*. Then, using the executing mechanism (actuator), this command is sent to the control object itself in the form of *U(t)*. Various hardware drivers, servos, stepper motors and other mechanisms can act as an executive mechanism.

Unlike the classical control scheme for voice control, the control device as a target receives not numerical parameters, but human speech or special sound signals. In this case, the control device must be supplemented with a special operator that translates the sound signal (human speech) into a command in the form of a machine command interpreter (figure 2).

At the same time, the task of recognizing voice command consists in constructing an operator for converting an audio signal (human speech) into commands in the form of a machine command interpreter. In a general case, such a command can be represented as follows: $K(p_1, p_2,..., p_n)$, where K is the kind of action or command, and $p_1, p_2,..., p_n$ are relevant arguments.

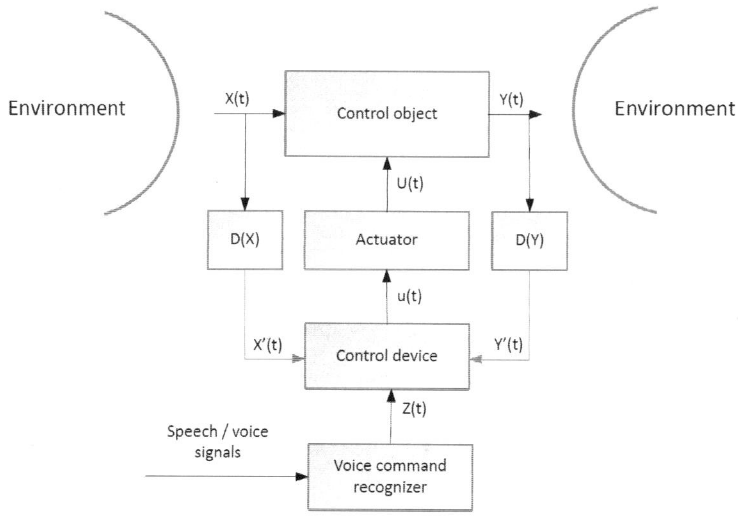

Figure 2. Convert speech to machine commands.

For example, the ***goto(x,y)*** command can be interpreted by the system as a change in the current position, while *(x,y)* is the vector defining the final desired position.

The main difficulty in recognizing voice command is the semantic interpretation of the words received and the assessment of fuzzy meanings. Let us consider the operation of the 'Voice command recognizer' block in more detail (figure 3).

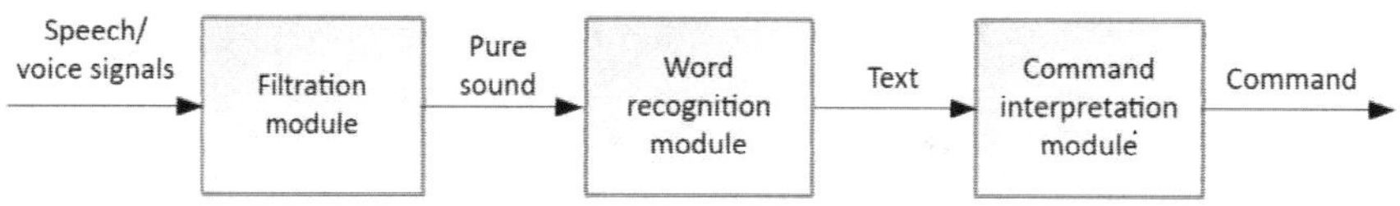

Figure 3. Module of the 'Voice command recognizer'.

An audio signal containing an indication of a person regarding the action that must be performed by the object arrives at the input of the module. At the first stage, speech is cleared of various noises using specialized sound filters. At the second stage, text *(speech-to-text)* is extracted from the received sound signal. The first two stages relate to a more general task, namely speech recognition, and have good industrial solutions in the form of software libraries. The third block is associated with the use of certain models and its interpretation in the form of a machine command. Currently, this task does not have an effective solution. In this paper, we propose a solution based on fuzzy calculations and the principle of expansion.

3. Proposed approach

One of the obvious solutions to the problem of forming a machine command based on a voice message is to set a direct correspondence between the received message and the action being performed, for example, in the form of a table. Let us consider the task of controlling the operation of the servo drive. In machine form, the command can be formed as ***go(p)***, where *p* is the parameter of the command, which can be interpreted as the duration of the command, the magnitude of the PWM signal that is supplied to the servo drive or some other interpretation. It is more convenient to set actions corresponding to voice messages in the form of a relationship table (table 1). For definiteness, we assume that p takes values in the interval [0,1].

Table 1. Table of relationships between voice messages and control parameter values.

Command / Control parameter	$p = 0$	$p = 0.5$	$p = 1$
go slowly	1	0	0
go medium	0	1	0
go fast	0	0	1

The disadvantage of this approach is the low expressiveness of such voice messages, and as a result, only rough control of the object. The expressiveness of correspondence tables can be increased using fuzzy calculations and the principle of expansion.

Table 1 shows the relationship between the set *go* = {“slowly”, “medium”, “fast”} and the set p = {0, 0.5, 1}. By introducing a fuzzy relationship, we can significantly expand the expressiveness of the voice command. The fuzzy ratio is calculated in three stages (figure 4).

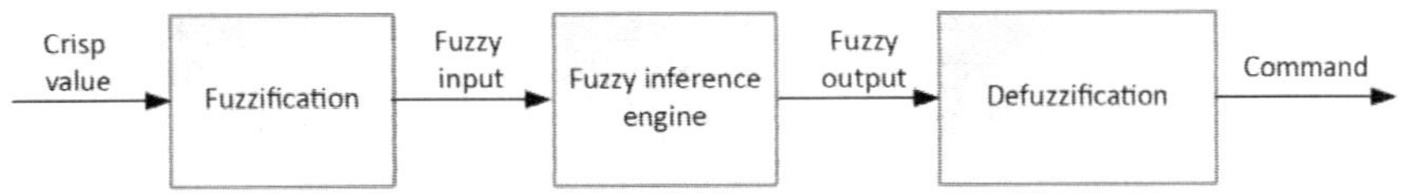

Figure 4. Fuzzy ratio calculation.

At the first stage, a voice message arrives. The voice message is decomposed into key terms related to the type of command and its parameters. Based on these key terms, fuzzification is performed, and as a result, the membership function for each command (in our case, for the element of the set ‘*go*’) is determined. The result is presented as a fuzzy number (1):

$$\frac{\mu_1}{x_1} + \frac{\mu_2}{x_2} + \ldots + \frac{\mu_n}{x_n} \tag{1}$$

where x_i is the i-th command, and μ_i is the i-th corresponding membership value.

On the basis of a table of crisp relationships, the principle of expansion allows a corresponding fuzzy relationship with more linguistic expressiveness to be built. For instance:

$$\frac{0.7}{slowly} + \frac{0.1}{medium} + \ldots + \frac{0.05}{fast} \tag{2}$$

The meaning of this fuzzy value can be described by the term ‘rather slowly’.

At the second stage, the fuzzy ratio is calculated. Let $x \in X, y \in Y$ be distinct sets. A relation R such that xRy is defined on these sets. On the set X, a fuzzy set A and the corresponding membership function $\mu_A(x)$ are defined, and on the set Y, the fuzzy set P and the membership function $\mu_P(y)$ are defined. Then, using the extension principle, we can construct a fuzzy relation of the sets A, P:

$$\mu_{aRp}(p) = S_{a \in A}[T(\mu_A(a), \mu_R(a, p))] \tag{3}$$

where T, S are the t-norm and t-conorm, respectively, for example $\max(.)$ and $\min(.)$. $\mu_A(a)$ is the membership function of the set A, and $\mu_R(a, p)$ is the membership of ratio R. Since a crisp relation is a special case of fuzzy relation, you can choose as a membership function the very crisp relation presented in table form.

Let us consider an example for the ***go(p)*** command. As a result of fuzzification, a fuzzy set A is obtained (4):

$$A = \left\{ \frac{0.7}{slowly} + \frac{0.1}{medium} + \ldots + \frac{0.05}{fast} \right\} \quad (4)$$

The fuzzy relation $\mu_R(a, p)$ is given (table 1). It is necessary to define a mapping of the set A to the set P.

As a result, we have:

$$B = \frac{\max(\min(0.7,1), \min(0.1,0), \min(0.05,0))}{p=0} + \frac{\max(\min(0.7,0), \min(0.1,1), \min(0.05,0))}{p=0.5} + \ldots + \frac{\max(\min(0.7,0), \min(0.1,0), \min(0.05,1))}{p=1} \quad (5)$$

Or (6):

$$B = \frac{0.7}{p=0} + \frac{0.1}{p=0.5} + \ldots + \frac{0.05}{p=1} \quad (6)$$

At the third stage, we form a control command. To retrieve the value of the control parameter, the defuzzification procedure is used. For example, the weighted sum:

$$p^* = \frac{\sum_i \mu_B(p_i) \cdot p_i}{\sum_i \mu_B(p_i)} \quad (7)$$

In our example (8):

$$p^* = \frac{0 \cdot 0.7 + 0.5 \cdot 0.1 + 1 \cdot 0.05}{0.7 + 0.1 + 0.05} = \frac{0.1}{0.85} \approx 0.12 \quad (8)$$

The main difficulty arises when performing fuzzification. It is difficult to build a universal procedure that allows you to efficiently extract membership values from a voice command when forming a fuzzy set A. One of the solutions to this problem is to set the corresponding memberships by listing (figure 5).

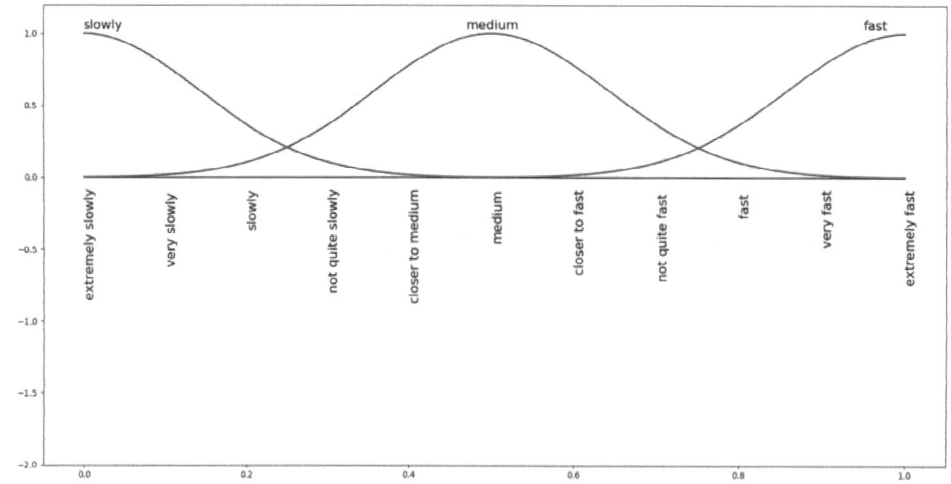

Figure 5. The fuzzification method of enumeration.

As a universal set, a virtual toggle switch (controller) acts here, the position of which can be set in the interval [0,1]. The approach using keywords allows you to set some discrete states of the toggle switch. The more such states there are, the more expressive the fuzzification procedure is. However, at the same time, it is more difficult to control the system because you need to keep in mind all the keywords and remember how they relate to each other. Ultimately, a person chooses the control of a system from a limited set. It is impossible to synthesize a new command for which a keyword is not provided.

This problem can be solved by using natural language analysis technologies. One of the most common is the bag-of-words technique. The essence of this approach is that all the normalized words known to the system are represented in the form of a binary vector. Each vector coordinate is associated with a specific word. If the word occurs in a phrase (voice command), then the coordinate takes the value 1, otherwise 0. Thus, each phrase (voice command) is translated into a vector of zeros and ones according to what words were used in the statement (phrase). Next, using the methods of machine learning, a model is constructed that produces a reaction to the received statement (in our case, this is the position of the virtual toggle switch).

Let us consider in more detail the fuzzification procedure using the bag-of-words technique (figure 6).

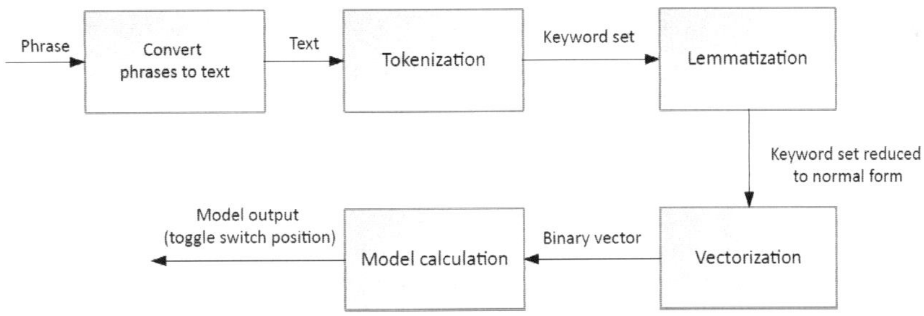

Figure 6. Fuzzification using the bag-of-words technique.

The input of the fuzzification procedure receives a certain statement (phrase). It is assumed that this statement will be in the form of an audio signal. At the first stage, the signal is translated into text. Next, using the tokenization procedure, interjections and prepositions are removed from the text. The result of this transformation is a set of keywords that are contained in the statement. At the stage of lemmatization, all keywords are brought into normal form, for example, nouns into the nominative case. Based on the words received and using the vectorization procedure, the text is translated into the feature space. The result of the operation is a vector of zeros and ones. The vector is fed to a model (for example, a neural network) which determines the position of the virtual toggle switch (taking a value from zero to one).

In order for the model to work effectively, it needs to be trained. The main difficulty here is that standard supervised learning is not possible. The loss function is calculated algorithmically, and the neural network affects the loss function through the control object. It is impossible to find the error gradient in this case.

Let us consider the training procedure in more detail. Fuzzification system training can be represented by the following stages:

1. Data collection and preparation
2. Setting the architecture and model parameters
3. Optimization of model parameters

Let us consider these steps in more detail.

Collection and preparation of data. In order to train the system, it is necessary to collect a sample where each change in the behaviour of the system will be associated with a user's voice command. To collect data, you must be able to work in a training mode with the system itself, or emulate the system. The system generates changes in its state, and the user comments on this change with the command that, in his opinion, describes this change. Once the data is collected, they need to be cleared. Inevitably, there will be inaccurate commands in the sample, outliers and erroneous comments. Data should be cleared and presented in the form of a table convenient for machine learning.

Setting the architecture and model parameters. It is proposed to take a neural network as a model. Neural networks have proven themselves in reinforcement learning and have flexible configuration tools. To train a neural network, it is necessary to choose an architecture and configure weighting coefficients.

Optimization of model parameters. During parameter optimization, the difference between the response of the system to a voice command and the behaviour of the system that has been commented on by the user is minimized. Let the sample *(X,y)* be collected during data collection, where *X* is the set of voice command converted using the bag-of-words technique, and *y* is the corresponding behaviour of the system. Then the optimization criterion can be represented as follows (9):

$$E = \frac{1}{N}\sum_{i=1}^{N}(S(NN(W, X_i)) - y_i)^2 \to \min_W \tag{9}$$

Here, X_i is the i-th voice command converted to a vector using the bag-of-words technique, y_i is the corresponding system change that the user commented on with the command X_i, $NN(W, X_i)$ is the reaction of the neural network to the voice command X_i when weight coefficients W. S are the response of the control object to the voice command X_i and its interpretation by the neural network. The difference between the response of the system and the desired response of the system is minimized by choosing the weights of the neural network. To optimize the weight coefficients, the differential evolution algorithm was used.

4. Experimental results

The proposed approach was tested on the next test task. The control object has *n* steps. At each step, the object changes its state. At the next step, the user is informed of the desired state of the object in the next step. The user executes a voice command, under the action of which the object changes its state. The task of the object control is to minimize the difference between the desired and actual behaviour of the object. As the control object, a fuzzy logic system was used with membership functions shown in figure 5, fuzzy relations (table 1) and defuzzification by formula (1). The system was controlled in two modes: 1) by keywords, 2) using a neural network and the bag-of-words technique.

In the process of data collection, the control object completely repeats the desired behaviour (figure 7). In this case, the user's voice command corresponding to the system transitions are recorded.

In control mode, at each step, the system under the action of a voice command is transferred from one state to another (figure 8).

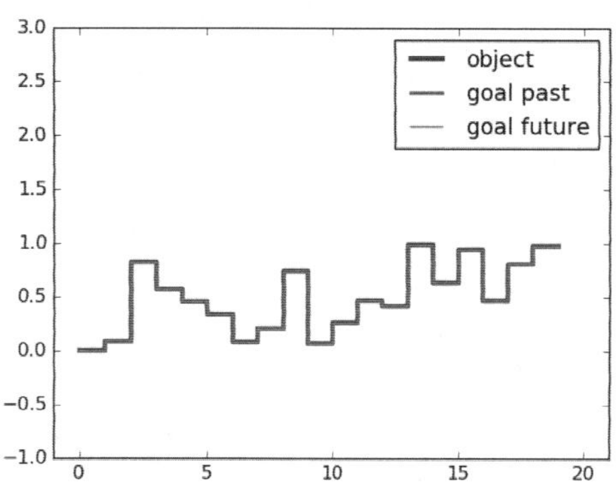

Figure 7. Data collection.

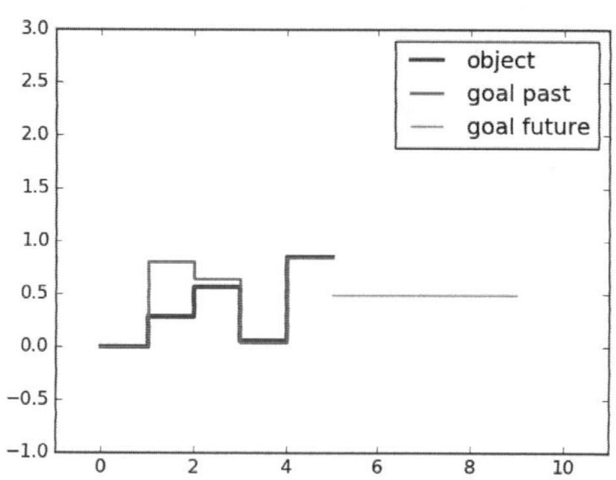

Figure 8. Example of object control using a neural network and the bag-of-words technique.

As can be seen from figure 8, at iterations 1-3, there are discrepancies between the desired behaviour of the object and the actual behaviour, and at iterations 3-5 the object practically repeats the desired state. The red line defines the desired behaviour of the object at the next iteration. According to the results of the system, the quality of control is determined in the sense of criterion E (9).

To compare the two approaches to control, a test solution of the problem was performed by each algorithm at 20 control iterations. For training the neural network, a sample of 200 examples was collected. After filtering the data, 191 values remained in the sample. For fuzzification, a neural network with two hidden layers of ten and five neurons, respectively, was used. To train the neural network, we used the differential evolution algorithm with the following parameters: 100 individuals per 100 iterations. The remaining parameters were taken by default. An example of a single solution to the problem is shown in figures 9 and 10.

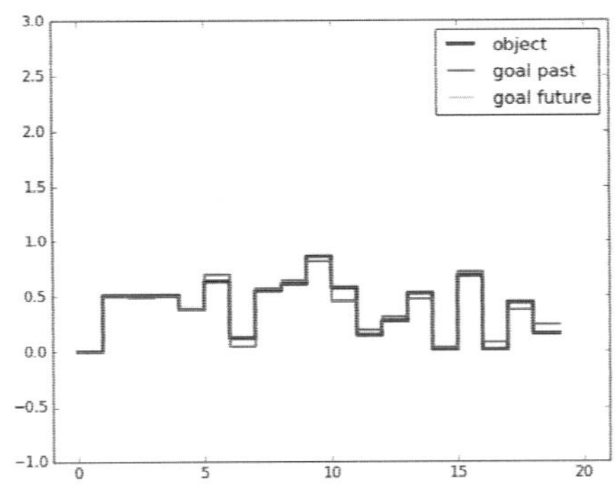

Figure 9. Control of the system using a neural network. The value of the criterion is 0.002523.

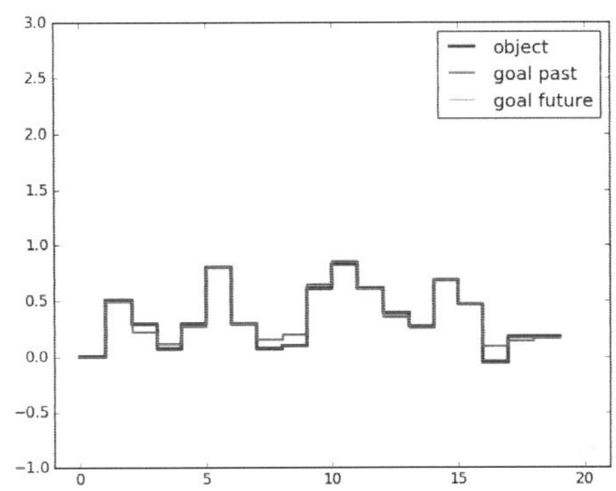

Figure 10. Control of the system by keywords. The value of the criterion is 0.002632.

Each task was solved by each approach 30 times. The results were averaged and are shown in table 2.

Table 2. Results of the proposed approaches.

Approach	FLS+NN	FLS+KW
E_min	0.000677	0.001204
E_max	0.007213	0.008832
E_mean	0.003173	0.003963
E_std	0.001606	0.001807

As can be seen from table 2, the statistical indicators for both methods are quite close. The main difference between the approaches was that the approach using the neural network is based on the household vocabulary of the operator, and the operator does not need much time to learn how to control the system. In the second approach, the system operator has to learn terms that are unusual for him, and a lot of time is spent on learning to control the system. The total number of keywords for the neural network approach was nine (table 3): 'medium', 'not very', 'a bit', 'same', 'a little', 'hard', 'very', 'confident', and 'slightly'.

For control by keywords, it was necessary to apply 11 commands (all the commands in figure 5 and the command 'same' or 'no change'). Moreover, in order to formulate 11 commands, it was required to use about 12 different words. The nature of system control is also different. When controlling the system using a neural network, the real behaviour of the object in many cases differs from the desired behaviour by a small amount. This leads to a deviation of the object from the desired behaviour. When controlling the system by keywords, the object has few cases of deviation from the desired state, but the deviation itself is more significant than when controlling it using a neural network. This can be explained by the fact that the operator controlling the system uses a lexicon that is not natural for him and from time to time makes serious mistakes in control. When controlling the system using a neural network, many small deviations can be explained by the low expressiveness of the operator's language during the initial emulation of the system at the data collection stage.

5. Conclusion

In this paper, an approach to the recognition of voice command using fuzzy logic systems is suggested. Moreover, two fuzzification methods are proposed: using keywords and using a neural network and the bag-of-words technique. It is shown that both of these approaches can be successfully applied to the task of recognizing voice command, while the approaches have differences. The first approach is based on an artificially created conceptual framework and can be difficult for the operator. The second approach is based on the natural language of the operator, but it requires the use of a certain procedure to build the conceptual framework and system training. Both approaches allow the necessary control accuracy to be achieved and should be applied taking into account the specified details.

At the same time, when controlling the system using a neural network, it is possible to increase the accuracy of control by increasing the expressiveness of the operator's natural language, for example, by repeating the procedure for collecting data and training the neural network. When controlling the system by keywords, the control accuracy can be improved by introducing additional keywords and with better (longer) preparation of the operator in terms of the established vocabulary.

Acknowledgements

This research is supported by the Russian Foundation for the Promotion of small forms of enterprises within project №14356GU/2019

References

[1] Trofimova A, Wiley T, Pagnucco M and Velonaki M 2019 Exploring Human Machine-Mediated Interaction for Applications in Social HRI. *In Proceedings of the 31st Australian Conference on Human-Computer-Interaction* 347-351

[2] Johannsen G 2009 Human-machine interaction. *Control Systems, Robotics and Automation* **21** 132-62

[3] Clark L, Doyle P, Garaialde D, Gilmartin E, Schlögl S, Edlund J [et al.] 2019 The State of Speech in HCI: Trends, Themes and Challenges. *Interacting with Computers* **31(4)** 349-371

[4] Paul R, Beniwal R, Kumar R and Saini R 2018 A Review on Speech Recognition Methods. *International Journal on Future Revolution in Computer Science & Communication Engineering,* **4(2)** 292-298

[5] Manish N and Reddy B J 2015 Review of voice control robot applications. *International Journal of Advanced Engineering* **1(9)** 671-674

[6] Rabiner L R 1989 A tutorial on hidden Markov models and selected applications in speech recognition. *Proceedings of the IEEE* **77(2)** 257-286

[7] Seong T W, Ibrahim M Z, Arshad N W B and Mulvaney D J 2018 A comparison of model validation techniques for audio-visual speech recognition. In *IT Convergence and Security* Springer, Singapore 112-119

[8] Liu Y Y, Li S, Li F, Song L and Rehg J M 2015 Efficient learning of continuous-time hidden markov models for disease progression. In *Advances in neural information processing systems* 3600-3608

[9] Mustafa M K, Allen T and Appiah K 2019 A comparative review of dynamic neural networks and hidden Markov model methods for mobile on-device speech recognition. *Neural Computing and Applications* **31(2)** 891-899

[10] Wijoyo S 2011 Speech recognition using linear predictive coding and artificial neural network for controlling movement of mobile robot. In: *Proceedings of 2011 International Conference on Information and Electronics Engineering (ICIEE 2011)* 28–29

[11] Eng G K and Ahmad A M 2005 Malay speech recognition using self-organizing map and multilayer perceptron. In *Proceedings of the postgraduate annual research seminar* 233

[12] Park J, Diehl F, Gales M J F, Tomalin M and Woodland P C 2009 Training and adapting MLP features for Arabic speech recognition. In *IEEE International Conference on Acoustics, Speech and Signal Processing* 4461-4464
[13] Hmad N and Allen T 2013 Echo State Networks for Arabic Phoneme Recognition. In *Proceedings of World Academy of Science, Engineering and Technology* **79**. World Academy of Science, Engineering and Technology (WASET) 424
[14] Seltzer M L, Yu D and Wang Y 2013 An investigation of deep neural networks for noise robust speech recognition. In *IEEE international conference on acoustics, speech and signal processing* 7398-7402

The effect of cation substitution on conductivity in the region of structural phase transition

L V Udod[1,2], M N Sitnicov[1], O N Bandurina[1] , and S O Konovalov[1]

[1]Reshetnev Siberian State University of Science and Technology, 31 Krasnoyarskii rabochii prospekt, Krasnoyarsk, 660037, Russia
[2]Kirensky Institute of Physics SB RAS, 50/38 Akademgorodok St., Krasnoyarsk, 660036, Russia

E-mail: luba@iph.krasn.ru

Abstract. The effect of non-stoichiometric substitution of chromium ions on the conductivity of bismuth stannates was studied. The measurement of the current-voltage characteristic in the temperature range 300-800 K and electric fields from -800 to 800 V/cm was carried out. A change of the conductivity type from hopping to tunnel-emission in the structural phase transition region is established. The influence of the concentration of the substituting ion on the change of the conductivity type was determined.

1. Introduction

Bismuth pyrostannate $Bi_2Sn_2O_7$ belongs to the pyrochlore family and undergoes structural transitions versus temperature [1]. The $\alpha \rightarrow \beta$ transition is realized in the region of 400 K and, $\beta \rightarrow \gamma$ at $T \sim 900$ K [2]. The $\alpha \rightarrow \beta$ transition has a ferroelectric nature with the noncentrosymmetry of the cubic structure, and the $\beta \rightarrow \gamma$ transition is a second-order transition. Recently, a low-temperature transition from a monoclinic structure to triclinic syngony at T = 140 K was detected [3]. The compound $Bi_2Sn_2O_7$ is dielectric and the substitution tin ions by 3d metals induces various types of exchange interactions. So, an antiferromagnetic exchange occurs between manganese ions, and at substitution by trivalent chromium, the spins of ions strive to line up ferromagnetically [4]. Replacement of Sn^{4+} by Cr^{3+} induces the charge ordering, and the interaction between the magnetic and ferroelectric subsystems, ferroelectric properties can be obtained as a result [5-7]. Variation of the ferroelectric properties by a magnetic field is realized using magnetoelastic interaction [8-11].

Substitution can lead to distortion of the crystal structure and a shift of phase transitions. For example, heterovalent substitution of Bi3 + and Sn4 + ions in Bi2Sn2O7 leads to a changing of the of the $\alpha \rightarrow \beta$ transition temperature [12].

The aim of this work is to study the effect of cation substitution of tin ions by chromium on the conductivity of $Bi_2Sn_2O_7$ in structural transitions region.

2. Experiment

The $Bi_2(Sn_{1-x}Cr_x)_2O_7$, where x = 0.1 and 0.15 was synthesized by the solid-phase reaction method according to the following scheme:

$$2(1\text{-}x)SnO_2 + 2xCr_2O_3 + Bi_2O_3 \rightarrow Bi_2(Sn_{1-x}Cr_x)_2O_7.$$

High-purity oxides Bi_2O_3, SnO_2, and Cr_2O_3 were used as initial materials. The initial mixture of the oxides pressed into pellets were placed in a furnace and held at temperatures from 700 to 950°C; the holding time was varied from 8 to 24 h. The powder X-ray diffraction pattern of samples was obtained at room temperature using a Bruker D8 ADVANCE diffractometer, a VANTEC linear detector, and the CuKα - radiation. All peaks, except several weak impurity peaks of an unknown phase corresponded, to the monoclinic Pc cell in the α-phase of $Bi_2Sn_2O_7$. The Rietveld refinement was realized using the TOPAS 4.2 program. The coordinates of all 176 atoms were fixed, since the number of 528 coordinates is comparable to the number of observed reflections. Nevertheless, even the fixed coordinates enabled us to correctly describe all the existing reflections, and the refinement resulted in low unreliability factors.

The measurements of electrical properties $Bi_2(Sn_{0.85}Cr_{0.15})_2O_7$ was carried out by four probe methods using an electrometer 6517V at 350 – 700 K.

3. Results and discussion

It is known that the conduction mechanism can be associated with Schottky and Poole – Frenkel emission, Nordheim – Fowler tunneling, and current limited by space charge [13].

The main conduction mechanism of $Bi_2(Sn_{1-x}Cr_x)_2O_7$, x = 0.1 and 0.15 at low temperatures (up to the structural transition α → β), where charge ordering is observed, is the model of currents limited by space charge (SCLC) with trap carriers. The voltage at which the linear region of I(U) curve begins, quadratically depends from the thickness of the sample. This proves the existence of SCLC mechanism controlled by shallow traps. In this region, the conductivity is described by Mott quadratic law [14]:

$$j = \frac{9}{8}\tau_\mu \sigma_0 \mu \frac{U^2}{L^3}, \quad (1)$$

where j is the current density, τ_μ is the Maxwell relaxation time, σ_0 is the electrical conductivity in the bulk of the material in the absence of charge carrier injection, μ is the mobility of the charge carriers, U is the applied voltage, L is the thickness of the sample.

The Poole-Frenkel conduction mechanism predominates at high temperatures (the β phase) [15]. A strong electric field applied to the sample changes the form of potential barriers for charge carriers between atoms of the crystal lattice. This leads to an increase of electrons in the sample due to overcoming the potential barrier. In this case, the current depends exponentially on the applied voltage and the exponent contains the square root of voltage:

$$I = e\mu n_0 \frac{U}{L} exp \frac{\beta U^{1/2}}{kTL^{1/2}}, \quad (2)$$

where, e is electron charge, μ is the mobility of the charge carriers n_0 is electron concentration in the conduction band in the absence of a field, k is the Boltzmann constant, β is the Pula-Frenkel constant, $\beta = \left(\frac{e^3}{\pi \varepsilon \varepsilon_0}\right)^{1/2}$, ε_0 is the electric constant, ε is the dielectric constant of semiconductor.

The linearity of the plots of curves I (U) in the Poole-Frenkel coordinates indicates that the contribution of charge carriers to the electric current arising on account the Poole-Frenkel emission. This means that the predominant charge carrier transport is due to both the hopping conduction mechanism and the tunnel emission of electrons.

In the region of the structural phase α → β transition in Bi2 (Sn1-xCrx) 2O7, x = 0.1 and 0.15, the conduction mechanism and the existence of inhomogeneous electronic states were determined from the current-voltage characteristics measured in an external electric field from -800 to 800 V/cm .

An increase of the substitution ion concentration leads to an increase of the conduction current. The shape of the current-voltage characteristics curves shows that in the region of T = 400 K Ohm's law is not satisfied (figure 1a).

Figure 1b shows the current – voltage characteristics in a logarithmic scale when a positive potential is applied to the samples of $Bi_2(Sn_{1-x}Cr_x)_2O_7$ compounds, x = 0.1 and 0.15, T = 400 K.

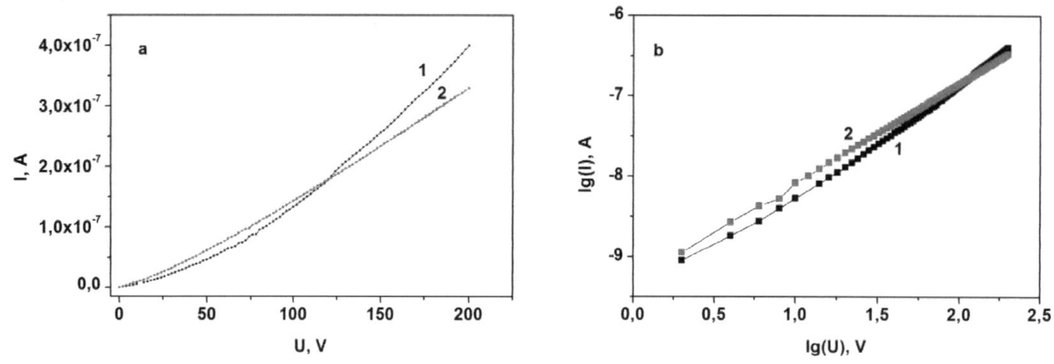

Figure 1. I–V characteristics of $Bi_2(Sn_{1-x}Cr_x)_2O_7$, x=0.1 and 0.15 at T=400 K, positive voltage. Curve 1 corresponds to x=0.1, 2 – x=0.15.

The current-voltage characteristic on a double logarithmic scale for bismuth pyrostannate x = 0.2 is linear and corresponds to $I \propto U^n$ $(n>1)$. This indicates that the transfer of charge carriers (holes) in $Bi_2(Sn_{1-x}Cr_x)_2O_7$, x = 0.1 and 0.15, is carried out by the mechanism of currents monopolar injection limited by space charges (SCLC) with the carrier charge traps.

To confirm this assumption, the dependences of currents from the square of the voltage were constructed, figure 2a. The curves are symmetrical for both negative and positive applied electric fields. Curve 1, which describes the I–V characteristics of $Bi_2(Sn_{1-x}Cr_x)_2O_7$, x = 0.1, is partially linear, which indicates the contribution of currents limited by the space charge. Curve 2 for x = 0.15 acquires a rectilinear shape in the field of high fields.

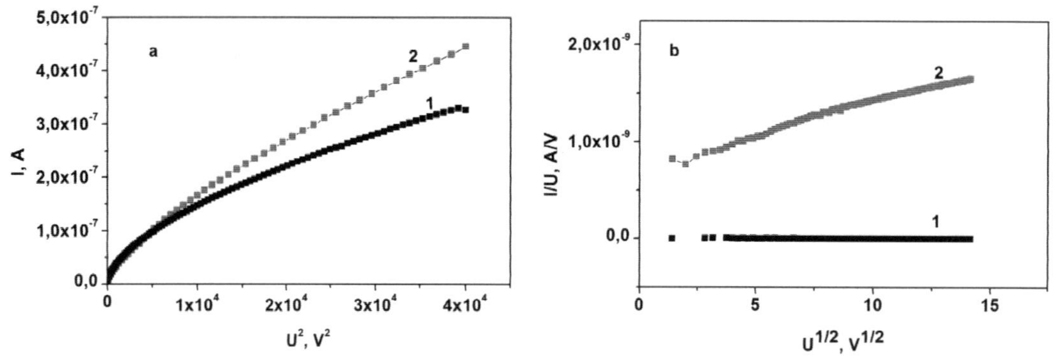

Figure 2. I–V characteristics of Bi2(Sn1-xCrx)2O7, x=0.1 and 0.15 at T=400 K, positive voltage. Curve 1 corresponds to x-0.1, 2 – x=0.15. (a) Dependence of the current from the squared applied voltage. (b) Pool–Frenkel I–V characteristic measured.

The dependences of the current strength from the applied voltage of positive polarity in the Poole-Frenkel representation for $Bi_2(Sn_{1-x}Cr_x)_2O_7$, x = 0.1 and 0.15 are shown in figure 2b. These dependences are symmetric for the positive and negative applied electric fields. Curve 1 corresponds to $Bi_2(Sn_{1-x}Cr_x)_2O_7$, x = 0.1, and tends to linearity on over the region of applied voltage. Curve 2 corresponded to the conductivity of the $Bi_2(Sn_{1-x}Cr_x)_2O_7$ compound, x = 0.15, where the linear region is observed in the high-field region. Thus, the conductivity will be described by equation (2) and corresponds to the Pool-Frenkel type for the sample x = 0.15.

4. Conclusions

In the region of the structural phase $\alpha \rightarrow \beta$ transition, the $Bi_2(Sn_{1-x}Cr_x)_2O_7$, x = 0.1 compound completely changes the conductivity type from hopping (Mott conductivity) to the Poole-Frenkel tunnel emission type. The conductivity of Bi2 (Sn1-xCrx) 2O7, x = 0.15 in the region of the $\alpha \rightarrow \beta$ transition, is mixed, part of domains carry out charge transfer according to the Mott type, and the other according to the Poole-Frenkel type. The ratio between these phases is approximately 50x50.

Acknowledgments
The reported study was funded by Grant of the President of the Russian Federation № MK-620.2021.1.2.

References

[1] Walsh, Graeme W Watson 2007 *Chem. Mater.* **19** 5158
[2] Shannon R D, Bierlein J D, Gillson J L, Jones G A and Sleight A W 1980 *J. Phys. Chem. Solids* **41** 117
[3] Udod L V, Aplesnin S S, Sitnikov M N, Romanova O B and Molokeev M N 2019 *J. Alloys and Compounds* **804** 281
[4] Aplesnin S S, Udod L V, Sitnikov M N, Eremin E V, Molokeev M S, Tarasova L S, Yanushkevich K I and Galyas A I 2015 *Phys. Sol. St.* **57** 1590
[5] Pal S, Banerjee A, Rozenberg E and Chaudhuri B K 2011 *J. Appl. Phys.* **89** 4955
[6] Efrem D'Sa J B C, Bhobe P A, Priolkar K R, Das A, Krishna P S R, Sarode P R and Prabhu R B 2004 *Pramana J. Phys.* **63** 227
[7] Fischer M, Malcherek T, Bismayer U 2008 *Phys. Rev. B* **78** 014 108
[8] Aplesnin S S 2000 *Phys. Rev. B* **61** 6780
[9] Applegate R, Hayre N R, Singh R R P, Lin T, Day A G R and Gingras M J P 2012 *Phys. Rev. Lett.* **109** 097205
[10] Kadowaki H, Ishii Y, Matsuhira Kand Hinatsu Y 2002 *Phys. Rev. B* **65** 144421
[11] Benton O, Sikora O and Shannon N 2012 *Phys. Rev. B* **86** 075154
[12] Aplesnin S S, Udod L V, Sitnikov M N, Kretinin V V, Molokeev M S and Mironova-Ulmane N 2018 *Mater. Res. Express* **5** 115202
[13] Brazis R, Pipinys P, Rimeika A and Lapeika V 1990 *J. Mater. Sci. Lett.* **9** 266
[14] Sze S M and Kwok K N 2007 Physics of Semiconductor Devices (Hoboken, New Jersey: Wiley) p 763
[15] Connell G A, Camphausen D L and Paul W 1972 Philos. Mag. **26** 541

IOP Conf. Series: Materials Science and Engineering 1230 (2022) 012022 doi:10.1088/1757-899X/1230/1/012022

New quality of measuring capabilities

E A Zhirnova, I V Trifanov, O A Sukhanova and D M Mednikov

Reshetnev Siberian State University of Science and Technology, 31 Krasnoyarskii rabochii prospekt, Krasnoyarsk, 660037, Russia

E-mail: sibgau-uks@mail.ru

Abstract. The issues of improving the quality of measurements based on the use of information technology and the informatization of the metrological infrastructure are investigated using the example of measurements of volumetric fluid flow. The functional diagram of an automated calibration unit for checking water flow meters is analyzed. As a result of the research, a model for the formation of a new quality of measuring capabilities has been developed. The use of metrological self-monitoring of calibration facilities is justified.

1. Introduction

Currently, scientific and technological progress presents new requirements for the accuracy and reliability of measurement information. The establishment and application of scientific and organizational foundations, technical means, rules and norms necessary to achieve unity and the required measurement accuracy is a difficult task, for the implementation of which regulatory requirements are developed to ensure the uniformity of measurements, accreditation criteria in a unified system of accreditation of legal entities and individual entrepreneurs to carry out work to ensure the uniformity of measurements [1, 2].

Since 2014, the Federal Agency for Technical Regulation and Metrology (hereinafter referred to as Rosstandart) has been actively pursuing an informatization policy aimed at providing the Russian economy, world-class metrological infrastructure by providing a qualitatively new opportunity to take into account information on the calibration of measuring instruments using modern information technology capabilities, increasing the effectiveness of state metrological supervision, and ensuring traceability in the framework of electronic certification of standards of units.

The introduction of resource-saving technologies, new electricity and water metering schemes leads to the development of improved calibration schemes, as well as the revision of the reference base. So, by order of Rosstandart on February 7, 2018 No. 256, a new state calibration scheme for measuring mass and volume of liquid in a stream, liquid volume and capacity for static measurements of mass and volume flow rates was approved. In accordance with this scheme, the portable calibration unit UPSJ 3 PM, used in the calibration of water flow meters, as standards of echelon 2, can now be certified only as standards of echelon 3 [3].

In this paper, we propose to consider the implementation of new measuring capabilities using the example of an automated portable testing apparatus UPPA designed for measuring, reproducing, storing, and transmitting units of the volume of liquid in a stream and volumetric flow rate of a liquid, as well as measuring the parameters of the surrounding and measured medium, time intervals.

IOP Conf. Series: Materials Science and Engineering 1230 (2022) 012022 doi:10.1088/1757-899X/1230/1/012022

2. Results and discussion

Installations are used for checking water meters at the place of their operation or in laboratory conditions in accordance with MI 1592-2015 and GOST 8.256-83 [4-7]. The principle of operation of the plants is based on measuring the liquid in the stream and the volumetric flow rate of the liquid using a primary flow element included in a single hydraulic circuit with a calibrated measuring instrument. The installation contains: a primary reference flow element (RFE), a measuring and computing complex (MCC), sensors of temperature, atmospheric pressure, and relative humidity of the environment, a temperature sensor of the measured medium, a control panel (CP), a functional diagram of the UPPA installation (figure 1).

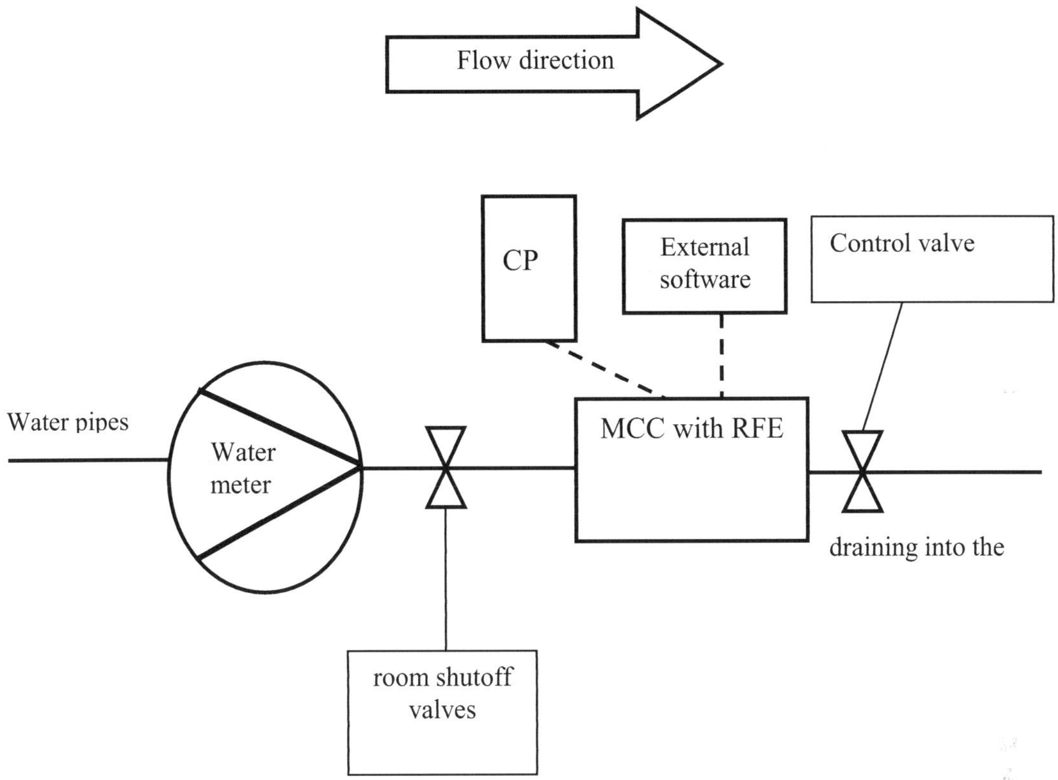

Figure 1. The functional scheme of the UPPA installation.

The units are connected to the hydraulic path in which the verified measuring instrument is located, using flexible hoses through quick-disconnect couplings. The fluid flows through the calibrated measuring instrument, the inlet hydraulic path, the primary flow element and merges through the outgoing hydraulic circuit with shut-off and control valves into the sewer or storage bank.

Volumetric flow rate and fluid volume are measured with a measuring and computing complex basing on data received from the primary flow element.

The measuring and computing complex controls the operation of the installation, automatically collects, processes and compares the received readings of the calibrated measuring instrument and the measuring instrument of the installation, as well as displays the received data on the installation display or via a Bluetooth communication interface to a peripheral device - a smartphone. Information from the calibrated measuring instrument is read out visually according to the readings of its indicator and entered into the measuring and computing complex, or the readings from the calibrated measuring instrument are marked by a fixation method using a smartphone with the Android operating system.

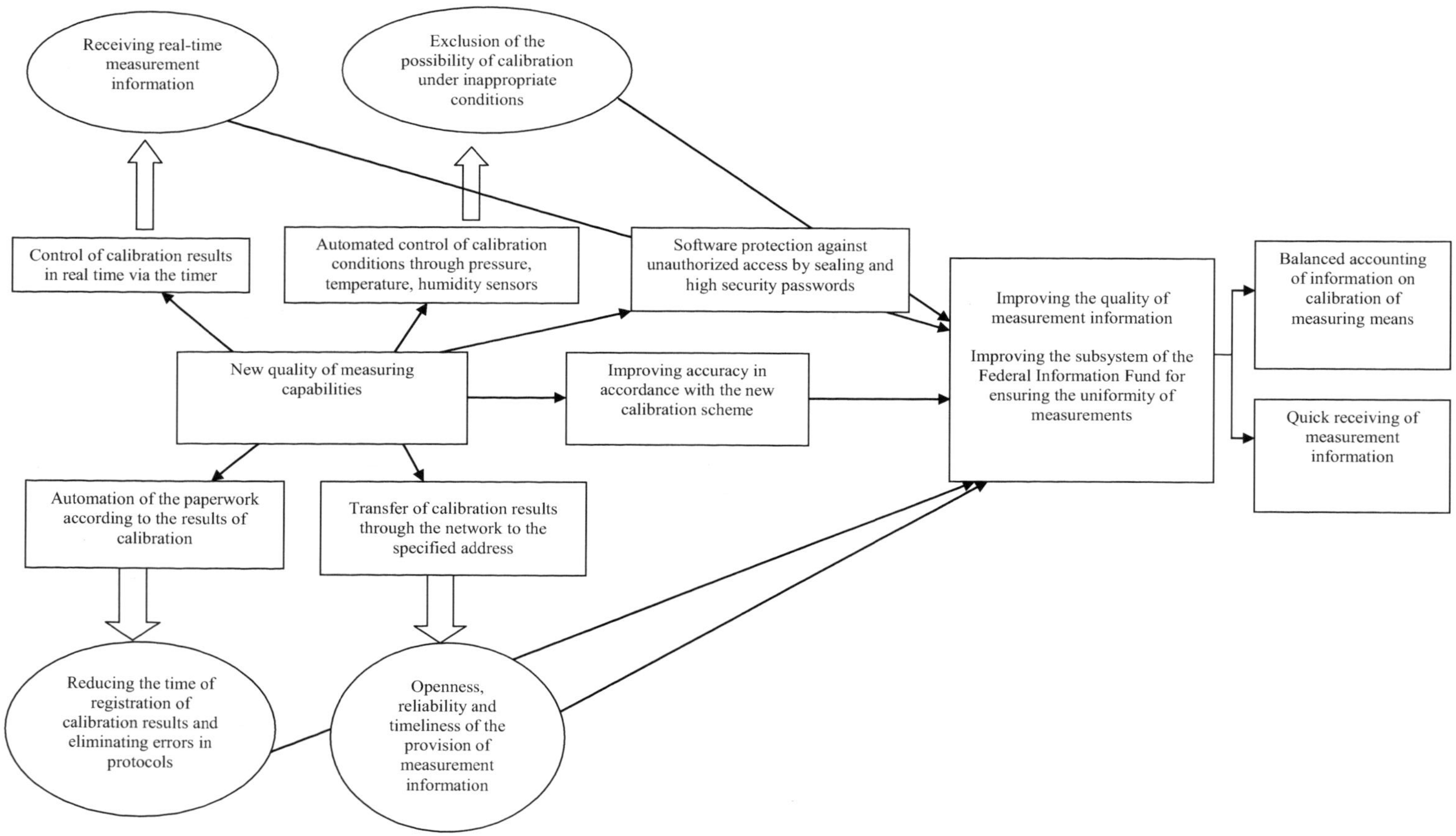

Figure 2. Model of a new quality of measuring capabilities based on information technology and informatization of metrological infrastructure.

An important advantage of the calibration installation UPPA, in comparison with other analogues [8-11], is the presence of a time sensor that allows you real time recording the results of calibration. After calibration of the measuring instrument in the volatile memory of the measuring and computing complex, the calibration protocol is saved, which can immediately go to the specified destination through the Internet. Installations have the ability to automatically register the environmental conditions and the measured medium.

In the course of the study, a model of a new quality of measuring capabilities was developed (figure 2), which clearly shows that the use of information technology and the informatization of the metrological infrastructure contribute to the creation of a new quality of measuring capabilities by automating calibration procedures, developing an integrated information database of calibration results, the possibility of online monitoring of completeness and reliability of calibration results.

3. Conclusion

Thus, basing on the theoretical studies:

- The issues of improving the quality of measurements on the example of measurements of the volumetric flow rate of the liquid were considered;
- The functional diagram of an automated calibration unit for checking water flow meters was analyzed;
- A model of a new quality of measuring capabilities has been developed, which is formed as a result of improving calibration schemes and implementing requirements for metrological infrastructure informatization, which ensures openness, reliability and timeliness of the provision of measurement information, as well as the use of automated control methods through integrated information networks, which serves as the basis for the introduction of resource-saving technologies;
- The use of metrological self-control of calibration facilities was justified.

References

[1] Federal Law №102 11.06.2008 *On ensuring the uniformity of measurements* GARANT system, available at: http://www.consultant.ru/document/cons_doc_LAW_77904/

[2] Order 30.05.2014 N 326 *On approval of the Accreditation Criteria, a list of documents confirming compliance of an applicant, an accredited person with accreditation criteria, and a list of documents in the field of standardization, the compliance with which is required by accredited applicants persons ensures their compliance with the accreditation criteria*" (with amendments on 08.19.2019) Ministry of Economic Development of Russia, available at: https://base.garant.ru/70712358/

[3] Order 07.02.2018 No 256 *On approval of the state verification scheme for measuring instruments for the mass and volume of fluid in a stream, fluid volume and capacity for static measurements, mass and volume flow rates of liquid* Federal Agency on Technical Regulating and Metrology, available at: https://www.garant.ru/products/ipo/prime/doc/71801798/

[4] GOST 8.156-83 *State system for ensuring uniformity of measurements. Cold water meters. Methods and means of calibration* (Introduced on 1983-03-01) (Moscow: Publishing house of standards) p 29

[5] Description of the type "Automated portable calibration units UPPA". Appendix to certificate No. 74965 on type approval of measuring instruments. available at: http://уппа.рф/images/Opisanie.pdf

[6] Automated portable calibration units UPPA. Manual. *Brant*, available at: http://уппа.рф/images/rukovodstvo.pdf

[7] MI 1592-2015 *Recommendation. GSI. Water meters. Calibration Method* (with amendments on 22.06.2018) (Kazan: FSUE VNIIMS) p 15

[8] Portable calibration unit. Manual. RKCP.407300.010 RE "VPU-Energo M" *TD-Rashodomer*, available at: http://www.td-rashodomer.ru/tech/tdocs/VPU-Jenergo_M-P_ustanovka_poverochnaja_perenosnaja_R%D0%95.pdf
[9] Portable calibration unit UPSJ 3 PM. Manual. U3PM.00.001 RE, *GC "New Technologies"*, available at: https://upsg.nt-rt.ru/images/manuals/manual_upsg_3pm.pdf
[10] Installation "Cascade-2P" Operation manual. *GC "New Technologies"*, available at: http://msk-call.ru/info/pover-ustanovki/104-kaskad-2p?format=pdf
[11] RMG 29-2013 *State system for ensuring the uniformity of measurements. Metrology. Basic terms and definitions* (introduced on 2015.01.01) (Moscow: Publishing house of standards) p 32

Increasing the autonomy of unmanned aerial vehicles by modifying the internal combustion engines used

A M Derben and S V Shelikhova

Reshetnev Siberian State University of Science and Technology, 31 Krasnoyarskii rabochii prospekt, Krasnoyarsk, 660037, Russia

E-mail: derben@sibsau.ru

Abstract. The most important characteristic of any unmanned aerial vehicles (UAVs) is the autonomy of operation. UAVs structures containing an internal combustion engine as an energy source usually use packaged units of general-purpose. The article proposes a modification of such engines, which allows improving fuel efficiency, reducing noise and exhausting temperature.

1. Introduction

Generally accepted direction for improving engine performance is the direction that uses engines with a modified duration of operating cycles (the Atkinson thermodynamic cycle).This cycle is characterized by a short stroke in the intake period and a long one in the power stroke [5]. However, the disadvantage of such units is the large size and high weight of the engine, low power, and high complexity of the unit as a whole [3]. A partial solution to all these problems is to simplify the design on the principle of the Miller cycle, using a conventional engine based on the classic scheme of the Otto engine [4]. In engines running on the Miller cycle, the intake valve closes much later than the intake cycle ends and part of the fuel-air mixture goes back to the intake manifold. Engines of this type are used, in particular, on Toyota Prius cars. Miller engines are also not without drawbacks: when the throttle is closed or partially open, pump losses significantly increase. This occurs when the fuel-air mixture passes through the partially open intake manifold twice: at the intake stroke and at the compression stroke, when the intake valve closes not immediately, but after a certain interval [4]. In this case, working at low rpm and high load is significantly difficult. The proposed modification of a conventional engine offers complete rejection of the throttle, and the amount of air is controlled electronically, through the premature closure of the air valve.

2. Modification project of a four-stroke general-purpose engine

Figure 1 shows the main idea of the modification. That is to reduce the combustion chamber in a conventional engine. In addition, the intake valve must be electronically adjustable and configured so that it closes before the end of the intake stroke, when the piston passes part of the distance from the upper dead center, a multiple of the reduction of the combustion chamber. The outlet pressure of a modern diesel engine is in the range of 3..4 Pa, for a gasoline engine is 2..3 Pa [2]. Taking this into account, it is obvious that in order to achieve optimal efficiency, it is necessary to reduce the combustion chamber and the intake stroke for a diesel engine in the range of three to four times, respectively, for a gasoline engine – from two to three times. The following example shows a gasoline engine with the chamber and intake stroke reduced by half.

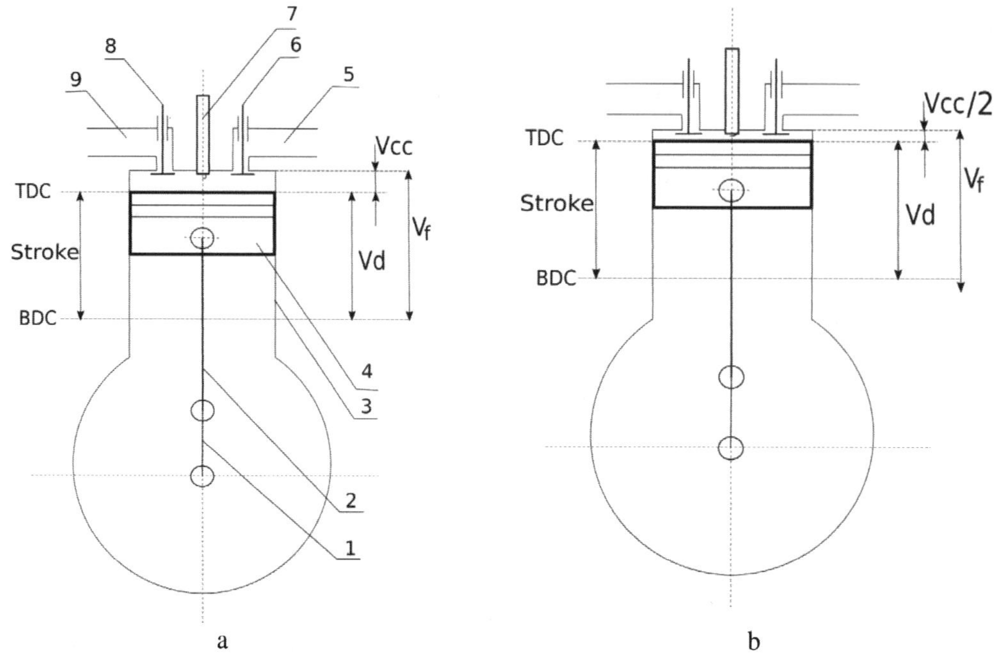

Figure 1. a – conventional engine, b – modified (1-crankshaft, 2-connecting rod, 3-engine housing, 4-piston, 5-exhaust manifold, 6-exhaust valve, 7-spark plug, 8-intake valve, 9-exhaust manifold).

3. Scheme of the conventional and upgraded engine operation

Figure 2 shows the initial phase of the engine intake cycle. This phase has no differences in the operation of both engines. The piston moves down. The intake valve is open. The fuel-air mixture enters the engine cylinders in the same way.

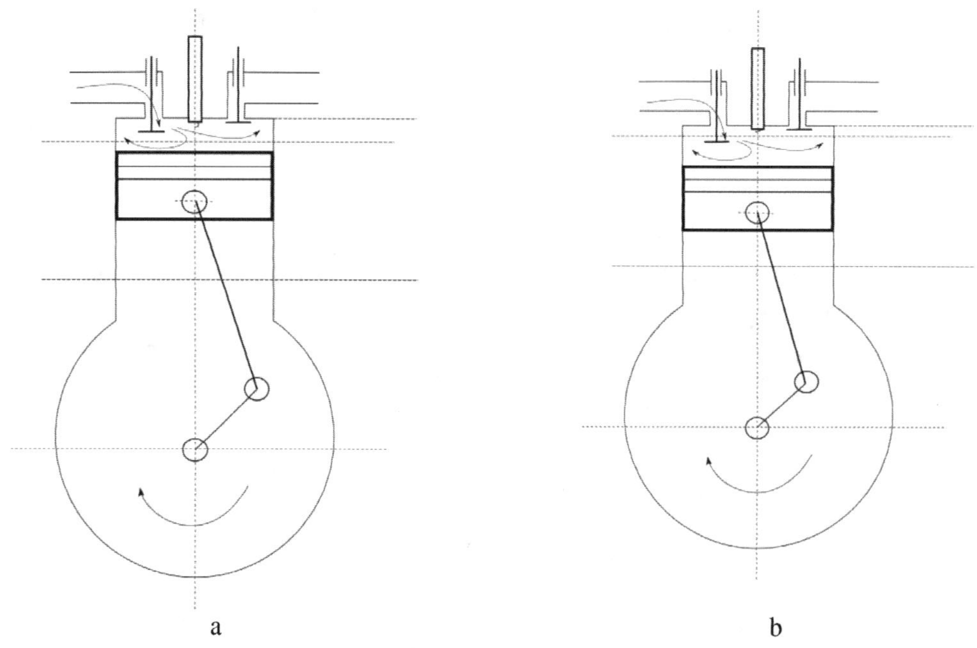

Figure 2. Start of the intake cycle (a – conventional engine, b – modified).

When the piston descends on a distance equal to the half of the power stroke, the intake valve is still open on a conventional engine, but on a modified one it is already closed. This condition is shown in figure 3. Note that to increase air filling at high rpm (over 4000 rpm) the valve should be closed as later as the higher the rpm. In this way, it is possible to solve the problem of all Atkinson-Miller engines that have insufficient dynamic range radically.

When the piston moves further down in the modified engine, an area of low pressure is formed in the cylinder and some part of the energy (usually the energy of the engine flywheel) is spent on overcoming the resistance of forces that pull the piston back into the cylinder.

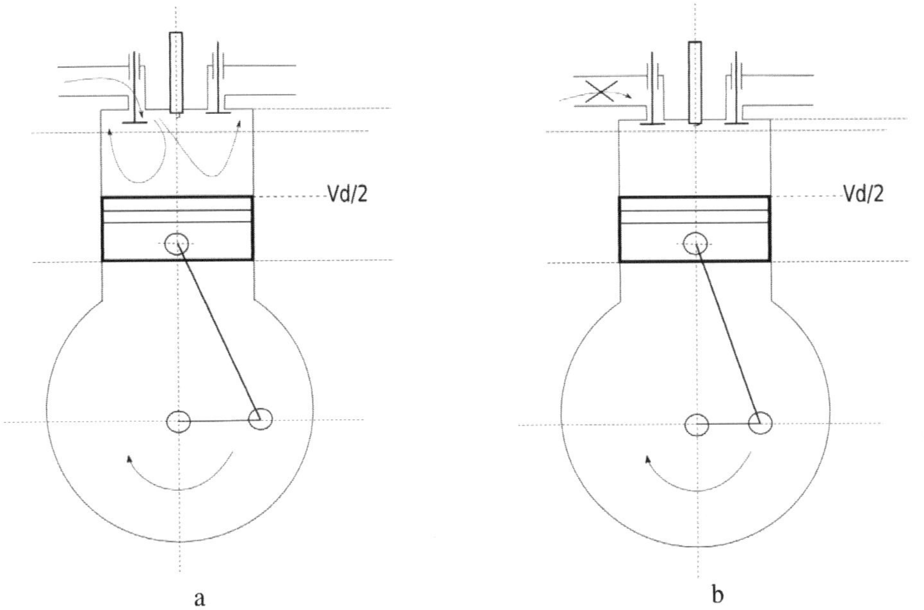

Figure 3. a – intake cycle (conventional engine), b – completion of the intake cycle (modified engine).

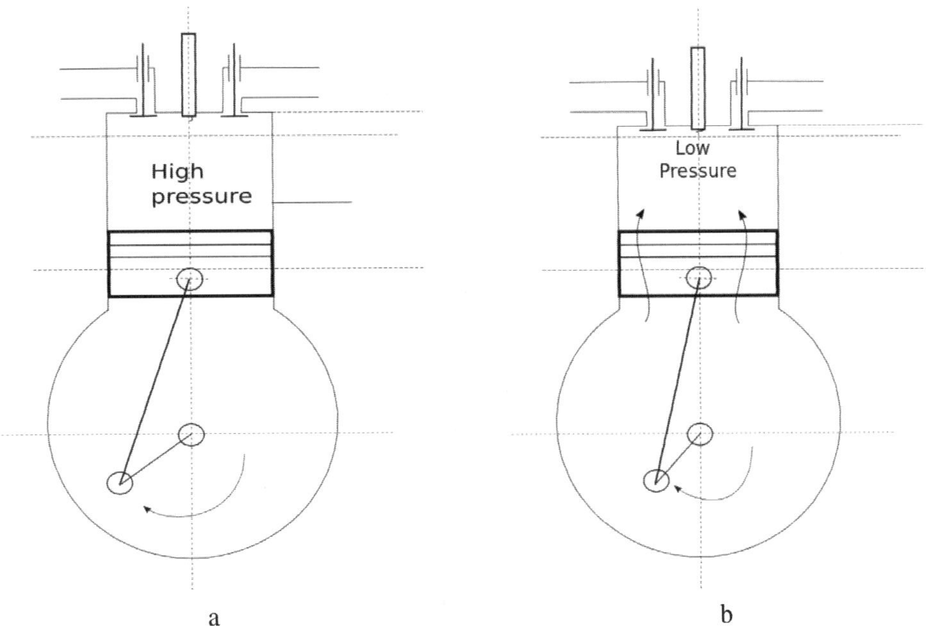

Figure 4. a – compression cycle (conventional engine), b – free running (modified engine).

At the bottom dead center (figure 4), for a conventional engine, the intake valve closes, the pressure increases, and the compression cycle of the mixture begins. In a modified engine, the low-pressure area pushes the piston up. Due to this force, free movement begins. As stated earlier, the modified engine does not have a throttle, which means that there is no pumping loss.

Figure 5 shows the moment when the stroke starts. In a modified engine, the pressure indicators correspond to those of a conventional engine [1], because the reduction in the volume of the compressible mixture is compensated by a smaller volume of the combustion chamber and the compression ratio is the same as that of a conventional engine.

The output stroke of a modified and conventional engine is completely similar.

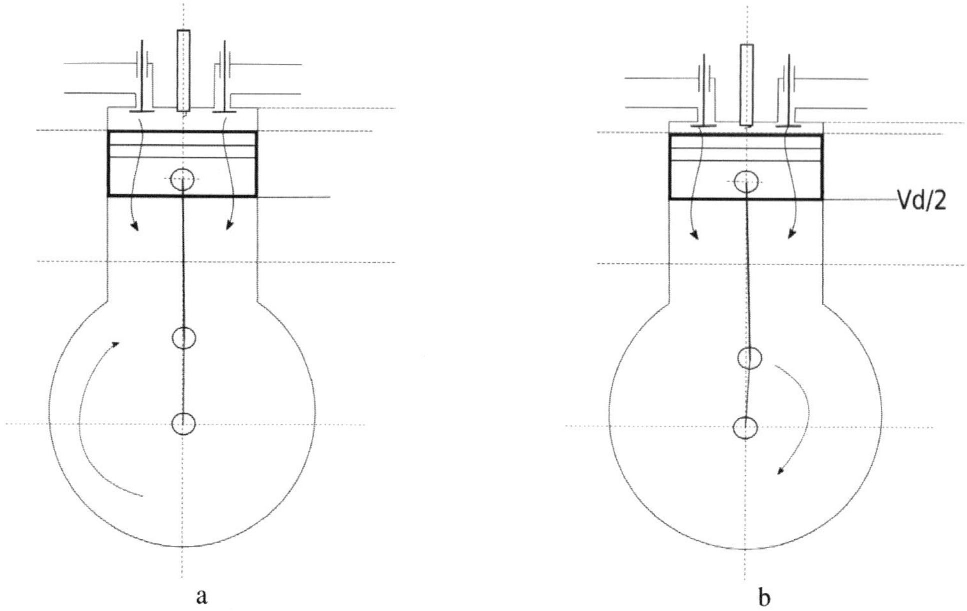

Figure 5. Power stroke, a – conventional engine, b – modified engine.

4. Indicator diagrams of the modified engine

Figure 6 shows a perfect diagram of the operation of the modified engine.

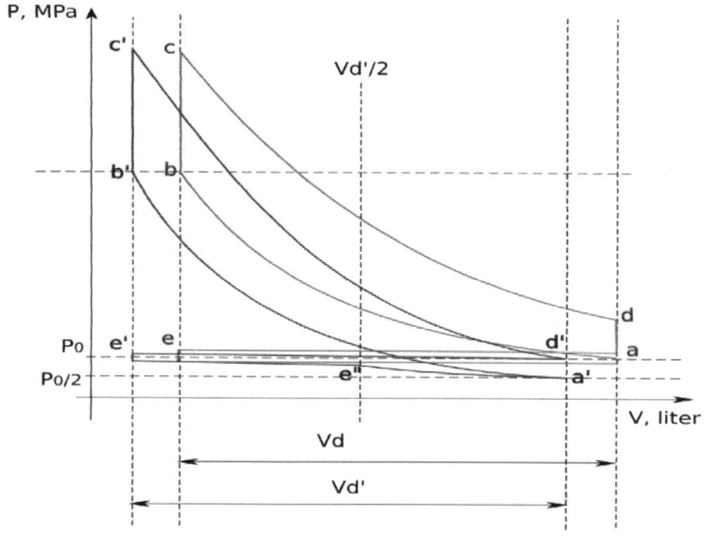

Figure 6. Indicator diagram: red – conventional engine, blue – modified engine.

Since the useful power of the unit is proportional to the area of the figure abcd for a conventional engine and the figure a'b'c'd' for a modified engine [2], when applying the resulting figures, one can see that the power of the modified engine will be lower by about 5..7 percent. This is due to the fact that the modified engine during the cycle of the power stroke at the same initial pressure has a half-smaller supply of fuel-air mixture, which also gives half the volume for the working body of the unit during combustion. This reduces the pressure at the end of the working stroke, and therefore leads to a small loss of power.

5. Conclusion

When using a modified engine according to the method described in this article, it is possible to achieve efficiency parameters of about 50% in all modes of throttling for gasoline engines and up to 60% for diesels. This will slightly reduce the output power of the unit. When designing engines of a modified type, it is advisable to increase the cylinder volume in proportion to this power decrease. The use of an electronically controlled electric valve can eliminate pump losses and raise the torque of the Atkinson-Miller engine at high rpm.

References

[1] Sharoglazov B A, Farafontov M F and Klementyev V V 2005 *Internal combustion Engines: theory, modeling and calculation of processes:* textbook on course "Theory of working processes and modeling of processes in internal combustion engines" (Chelyabinsk: Publishing House. SUSU) p 403.

[2] Makushev Yu P, Kanya V A, Litvinov P V and Voitenkov S S 2017 *Dynamics of internal combustion engines: educational- methodological guide* (Omsk: SibADI), available at: http://bek.sibadi.org/fulltext/esd363.pdf.

[3] Taku Tsujimura and Yasumasa Suzuki 2019 Development of a large-sized direct injection hydrogen engine for a stationary power generator *International Journal of Hydrogen Energy* (Japan) **44** 22 pp 11355-11369

[4] Rahim Ebrahimi 2018 A new design method for maximizing the work output of cycles in reciprocating internal combustion engines *Energy Conversion and Management* (Iran) **172** pp 164-172

[5] Gęca M S, Rybak A and Hunicz J 2018 A simulation study into the Atkinson cycle engine utilizing adjustable crank mechanism *IOP Conf. Series: Materials Science and Engineering* **421** 042021

Engineering of GIS for environmental monitoring of mining landscapes

Y P Yuronen[1], A V Bakhtina[1], E A Yuronen[1,2]

[1]Reshetnev Siberian State University of Science and Technology, 31 Krasnoyarsky Rabochy Av., 660037 Krasnoyarsk, Russia
[2]Siberian Federal University, 79 Svobodny Av., 660041 Krasnoyarsk, Russia

E-mail: anna-denisyuk@yandex.ru

Abstract. Features of GIS design, designed to monitor the state of disturbed lands at open mining facilities, natural and climatic conditions in the mining area, are considered in order to inform the adoption of administrative and management decisions. The introduction of such systems will make it possible to carry out operational environmental monitoring of lands subject to technological violations, reduce the time for preparing reporting and technical documentation when designing and conducting measures aimed at land reclamation, make decisions on the technology and timing of reclamation of subsoil use facilities, and also reduce the time for making such decisions.

1. Introduction

In view of the rapid development of industry, it is becoming increasingly important to assess the man-made impact of human activities, especially in places of mining, building materials and peat, as well as in places of their enrichment and processing [1]. Open mining activities, which disturb natural ecosystems, cause significant damage to the natural environment. The area of disturbed land by coal mines as of 2016-2017 years amounted to about 180 thousand hectares [2], iron ore quarries – more than 43 thousand hectares [3], in addition, non-mineral deposits are being developed.

Despite the obligation to restore ecosystems on disturbed lands, reclamation work in most cases is not carried out on time, not in full, leaving large areas for self-restoration. The climate, hydrological regime, composition of overburden can lead to the effect of a slow rate of land restoration, so there is a high probability that the restoration process will go on for a long period of time or will not occur at all. Recultivation of disturbed land is carried out with the aim of further use in forest, agriculture, water farms, for construction, recreational, environmental and sanitation measures. It is not possible to assess the effectiveness of the work performed and determine the prospects for the further use of such lands without comprehensive monitoring studies. Studies of the ecological status and level of restoration of natural components on mining landscapes are carried out effectively on the basis of analysis of remote sensing data [1, 2].

2. Objectives and opportunities of the projected GIS

The projected geoinformation system is designed to provide information on the state of disturbed lands at open mining facilities, natural and climatic conditions in the mining area, with the aim of informational support for the adoption of administrative and management decisions.

The projected opportunities:

1) record keeping of registered users in the system record keeping of registered users in the system;
2) 24-hour access to the system;
3) maintenance of structural parts of mining enterprises geobase, their condition throughout the life cycle, as well as elements of assessment of natural and climatic characteristics of the region;
4) providing access to the necessary information and functionality to the user for making a decision on reclamation, drawing up land restoration forecasts;
5) implement interactions with other data sources necessary for decision-making;
6) preparation of reporting documents in the form of graphs, charts, maps and reports on the results of the carried out reclamation, rate of vegetation restoration, statistics on administrative units and other indicators depending on user requests.

The designed GIS should provide access to information through the Internet through a web interface. Input data loaded by the geodatabase administrator can be represented by raster and vector models, the attribute table must meet the requirements of the subject area. The output is user-set scale maps, reports, and charts based on user requests.

In accordance with the subject area, purpose and tasks, the following thematic layers can be identified: administrative-territorial division; relief; hydrography; road routes; hydrometeorological data; geology of the monitoring area; disturbed land.

3. Design of GIS

On the basis of the selected thematic layers, the geodatabase is designed, which is fully integrated into the GIS of environmental monitoring of mining landscapes.

The projected geographic information system will be presented to users as a geoportal based on the multi-user client-server architecture (figure 1)

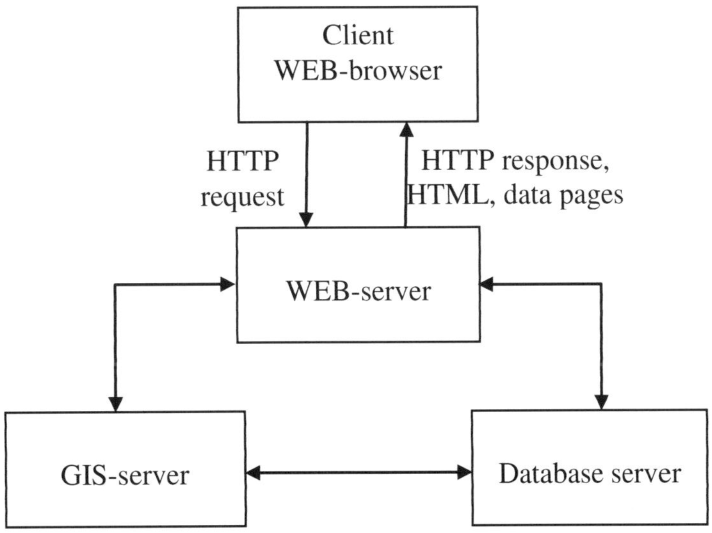

Figure 1. GIS architecture of ecological monitoring of mining landscapes.

An independent model of client interaction with the Web server may be used to improve GIS efficiency. Each operation will be performed asynchronously in a separate thread, allowing the client to perform other actions at that time.

A conceptual model has been developed to represent the geodatabase structure (figure 2), reflecting the internal relationships. A distinctive feature of the geodatabase from databases is that tables do not always have a tabular relationship by primary code, often attribute tables of spatial objects have a spatial relationship with other attribute tables. For example, by means of a spatial query, the user can find all enterprises located in the region. The development of a conceptual model of the geodatabase determines which data is related to each other. Thus, by defining these links, the user can search for objects from one table by the attributes of another table.

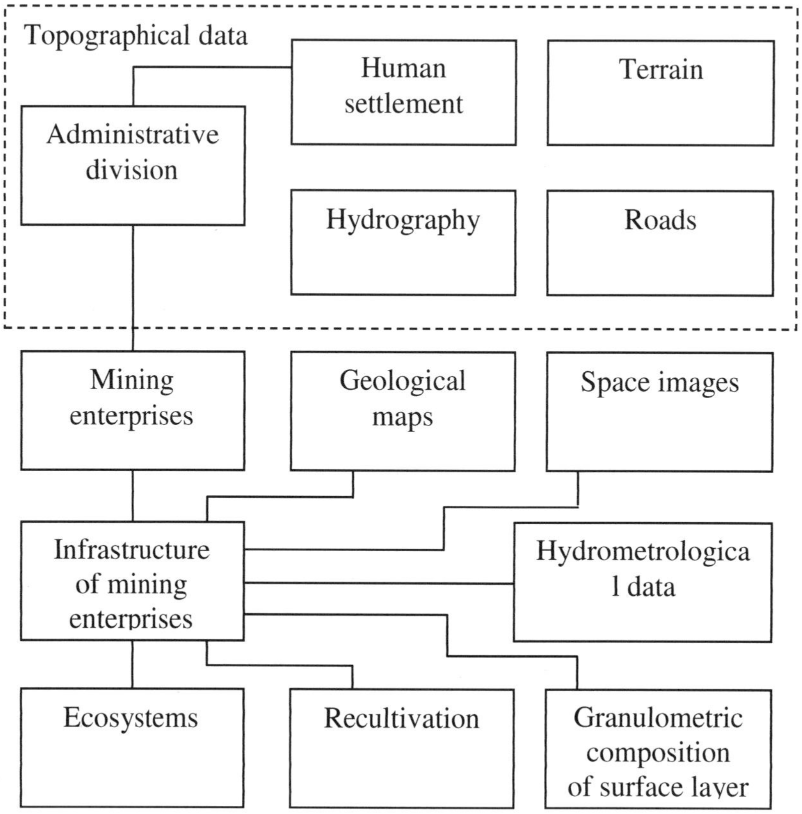

Figure 2. Geodatabase conceptual model.

The next step is logical design. The purpose of this step is to convert a conceptual model based on the selected data model into a logical model.

The structure of the tables and the attributes of the coatings must be defined to develop a logical geodatabase model. Based on the developed conceptual model and structure of attribute data tables, a logical model can be developed. (figure 3).

Thus, this GIS will allow to remotely receive information through Internet access using a web browser, export materials in the form of maps, reports, charts and graphs. For the further implementation of GIS, functional and non-functional requirements were formulated, a geodata database was designed, a system for encoding and classifying objects and their attributes in accordance with the subject area was developed.

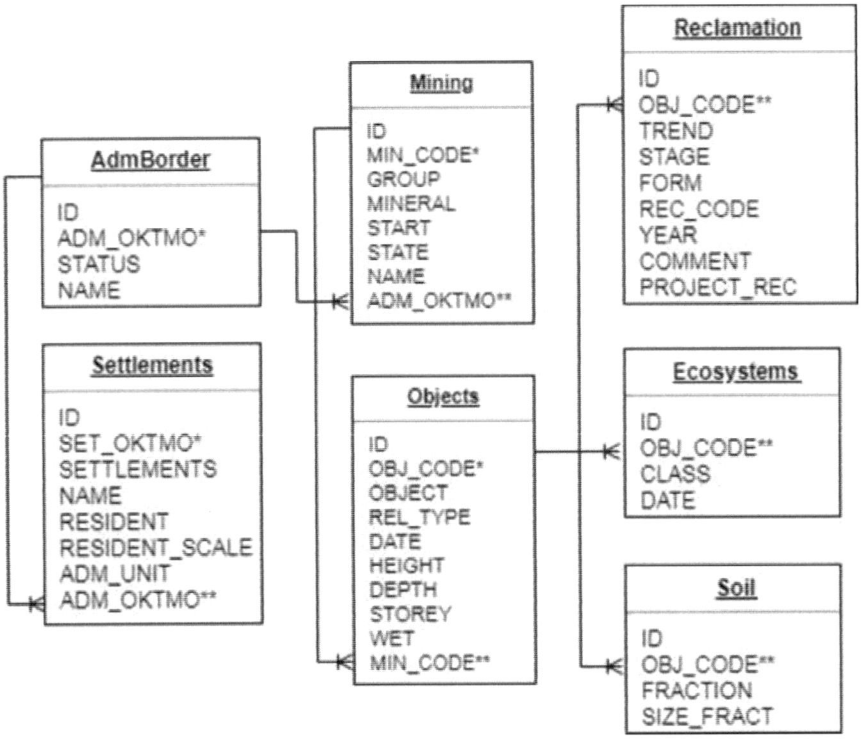

Figure 3. ER-model of geodatabase tables linked by primary and secondary codes.

As a further development of the system, one can consider: the addition of new layers of objects of disturbed land and other characteristics that allow more detailed analysis of the ecological state of disturbed land.

4. Conclusion

In general, the introduction of this GIS will reduce the time for making administrative and managerial decisions, carry out operational environmental monitoring of lands subject to technological violations and make decisions on the technology and timing of reclamation of subsoil use facilities. In addition, GIS will reduce the time to prepare reporting and technical documentation for the design and implementation of measures aimed at land reclamation.

References

[1] Rodaeva V V *Restoration of vegetation cover on dumps of brown-coal deposits of Southern Primorye:* abstract of the dissertation of the candidate of biological sciences (USPU: Ussuriysk)

[2] Zenkov I V *et al.* 2017 *Coal mines of Russia from space. Mining and ecology of disturbed lands:* monograph (Krasnoyarsk: SFU) p 528

[3] Zenkov I V *et al.*2018 *Iron ore quarries of Russia from space. Mining and ecology of disturbed lands:* monograph (Krasnoyarsk: SFU) p 664

Observations on the construction methods used in Soviet-era shore navigational aids in the northern reaches of the Yenisei River

A E Goncharov, A A Gorodov

Reshetnev Siberian State University of Science and Technology
31 Krasnoyarskii rabochii prospekt, Krasnoyarsk, 660037, Russia

E-mail: goncharovae@sibsau.ru

Abstract. Intense navigation in the northernmost reaches of the Yenisei River during the Soviet era produced a rich industrial heritage. Among the more conspicuous objects are the remains of shore navigational aids, particularly leading lights and beacon towers. The surviving structures offer insights into the navigation methods used in this part of the Arctic. Navigational aides were also among the last industrial structures in the Soviet Arctic to be built of wood. Besides, they are important examples of structural engineering in the Arctic, meaning that they were designed to cope with the harsh environment.

1. Introduction. The Yenisei River is a natural riverine continuation of the Northern Sea Route (NSR), which links the Arctic Ocean to the interior of Central Siberia. The river is continuously navigable for seagoing vessels to its only surviving seaport – Dudinka (69°24′N 86°11′E). For ships drawing under three meters, the head of navigation is the regional capital city – Krasnoyarsk situated over 2,500 km inland. This unique geographical feature of the Yenisei has made it an attractive transportation corridor since the 1870s, when the first seagoing steamers ascended the river for the purpose of delivering European industrial goods to Siberia. Thenceforth, thousands of vessels have navigated the vast stretches of the Yenisei making it the second busiest river shipping lane in Siberia. Thus, in 2014, vessels operating in the Ob' and Irtysh basins transported over twelve million tons of cargo, while only approximately three and a half million tons were carried up and down the Yenisei and its tributaries. Meanwhile, the average distance this cargo was carried comprised 550 km for the Ob' and Irtysh system, while 1,370 km was the figure for the Yenisei, making navigation distances here significantly greater [1]. Today, the intense development of oil mining on the Taimyr Peninsula has put additional strain on the riverine transportation system.

Before the arrival of modern satellite navigation systems, shipmasters relied on a system of lateral navigation aids. The history of this system in connection to Yenisei Arctic shipping marked a centennial anniversary in 2020. Its usage continues to this day in the upper and middle reaches of the river. However in the northernmost reaches, where the river widens to several (and even tens of) kilometers, the lateral system aids are now largely disused. While some beacons and leading lights towers (especially steel truss structures) continue to operate, timber towers have largely been abandoned and left to deteriorate in the harsh Arctic climate. Here it is important to note that during the last years with the increase of air temperatures in the Arctic regions followed by the subsequent melting of permafrost layers, charges in precipitation rates, and increased activeness of microorganisms threatens the existence of preserved wooden historical structures [2]. As these sites are under immediate threat, it is important to develop a precise strategy for preserving existing industrial heritage sites located in the Arctic. Employing structural analysis provides a way to better understand original construction methods and as such is considered pivotal to the preservation and restoration of historical structures [3].

The timber leading lights towers, which dominate the landscape of the Arctic reaches of the Yenisei (north of 68°N), were built during the Soviet epoch. Some remains can be dated back to the very first beacons built during the 1920s and 1930s. They are tangible fragments of evidence speaking of the technological development of the Arctic navigation system (both maritime and riverine), paralleling the history of the sea route itself. As such, they provide a firm basis for the need to study structural evidence in the northern polar region.

In view of the aforementioned, the study of past navigation facilities should become a subject for a comprehensive study on the industrial archaeology of the Northern Sea Route. Due to the immensity of this task, in this paper, we shall narrow our focus to the structural features of leading lights towers, which remain the most distinctive anthropogenic structures on the riverbanks and islands of the lower Yenisei. This study is a presentation of a preliminary investigation of these structures, built on field data collected by us in 2019.

Internationally, there has been but a handful of studied dedicated to historical navigational aids, among which is a descriptive essay on aids in the Mississippi River mouth [4]. There is also a paper on the history of providing safe navigation for vessels operating on the river Ob' and its tributaries during the Second World War [5], however, as with other related studies, it is based on documentary, rather than material evidence.

2. Field data, collected in 2019 and its analysis.

In 2019 during a project hosted by Reshetnev University (Krasnoyarsk, Russia), a team of researchers headed by the lead author of this paper transected a two-way water route along the right bank of the river Yenisei from Dudinka to cape [*mys*[1]] Muksuninskii (70°18'1"N 82°59'8"E) (Fig. 1). The usage of inflatable smallcraft allowed the team to investigate a number of historical sites, the majority of which are valuable as industrial heritage and industrial archaeological monuments. As with many industrial objects, these have largely remained neglected by professional researchers, archaeologists and historians alike. The latter dismiss such objects as being outside the scope of archaeological science, whereas the former seldom undertake extensive fieldtrips, preferring to conduct documentary research in archives.

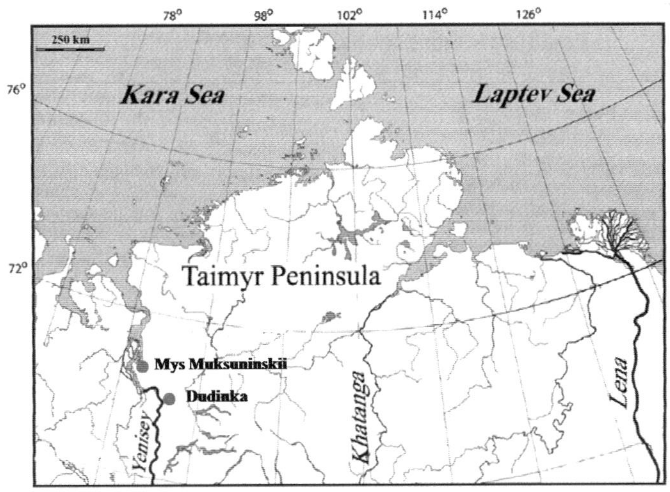

Fig. 1. Map of the Taimyr (Taymyr) Peninsula showing the study area.

[1] The present authors are responsible for all translation and transliteration on this text.

The expedition gathered data on 18 sites, including seven settlements, four cemeteries, and three leading lights towers. During the fieldwork, we were able to recover structural evidence on many of the towers, particularly those fully intact or with standing remains.

Fig. 2. Leading lights tower at Mys Krestovskii (c. 1950s)

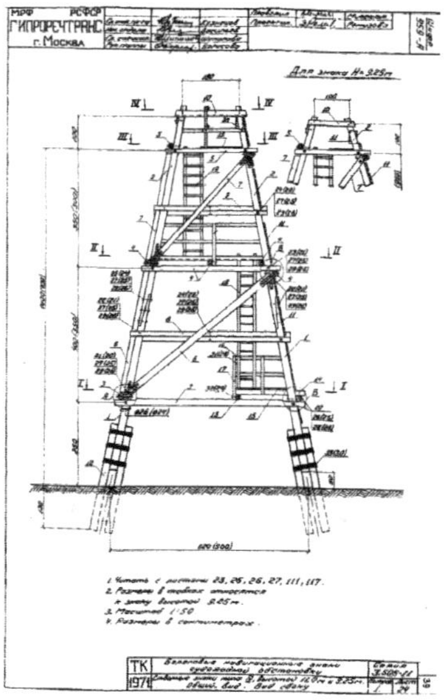

Fig. 3. Scheme of standard 9.25 m in height tower for coastal areas.

Pairs of leading (range) marks are elemental navigational aids positioned in such a manner to be aligned with one another in order for the navigator to follow the leading line on the same bearing. Following a framework of bearings, the ship is guided through the deepwater channel and kept clear of submarine obstacles. Visibility is a key factor in the design of this type of aid. In monotonous landscapes such as the Siberian tundra, any freestanding structure is highly visible. The addition of a painted (typically, black and white) dayboard to the observable side of the tower further enhances the visibility of the towers. For autumn and, from the mid-1970s, winter navigation the towers were fitted with electrical lights powered by alkaline batteries making them visible during the polar night.

Fortunately, because the construction and maintenance of the structures had been staged by centralized state organizations, we have precise and detailed data on how they were built. Preserved technical documentation enables us to conduct a detailed analysis of the structures and compare them to preserved examples. For instance in Fig. 2, the standing truss tower at mys Krestovskii is perfectly comparable with a plan for a similar construction.

As with all industrial structures built in the Soviet Union, the construction of the towers was regulated by a series of state-issued standards, known as *GOST*(s). One of the latest of these documents was the GOST 16150-71 *Znaki navigatsionnye beregovye. Tipy, osnovnye razmery i tekhnicheskie trebovaniia* [Shore Navigation Signs: Types, Dimensions, and Technical Specifications] issued in 1971. This standard, in turn, provided common layout for producing specialized technical documentation such as the 1972 Project for Wooden Shore Navigation Signs (the 3.505-11 Series) by the *Giprorechtrans* [State Institute for Projecting Riverine Transportation]. Thereby the construction of navigation towers as well as other similar structures was regulated by a series of technical documents; thereby, standardizing them.

In the table below are shown the main structural elements and their dimensions of the 9.25 m tower.

Component	**Material**	**Diameter in cm or grade**	**Length in cm**	**Unit of measurement**	**Amount**
Vertical support	Pine	24	500	Number per m^3	4/1.08

Vertical support	Pine	14	700	4/0.60
Collar beam	Pine	20	600	4/0.92
Collar beam	Pine	16	440	7/0.73
Collar beam	Pine	16	300	3/0.33
Knee brace beam	Pine	20	600	3/0.69
Knee brace beam	Pine	16	460	3/0.34
Diagonal brace	Pine	14	730	2/0.32
Diagonal brace	Pine	14	520	2/2.20
Brace	Pine	15x2.5	230	3/0.03
Planks for dayboard	Pine	20x2.5	unspecified m^3	0.76

The choice of wood as the primary material for constructing the towers was dictated both by the cheapness of this material and its availability. Besides, it was expected that these structures were to be built and maintained by riverine and maritime facilities without the need for special factories or personnel [6]. Furthermore, as seen from the choice of timber type, the tower had to be relatively lightweight so to be easily transported to the building site, possibly using manual labor. The prefabricated structure was delivered to the building site, where it was assembled on the ground as two separate trusses. These were then erected in succession using a set of winches and set into pre-excavated holes. The following procedure involved joining the two trusses with braces and building a four to five decks and ladders for subsequent tower maintenance.

An important feature of wood to consider when designing structures having a potential buckling instability is its orthotropic properties. It is known that wood is better at resisting loads such as compression, shear and tensile stress in the direction of the fiber growth rather than in transversal directions [7]. Due to this the towers were constructed as three-dimensional lattice rigid structures. Lattice towers are known to be very effective at withstanding both dynamic and static loads, notably those from the environment [8]. The greatest loads for the wooden towers in the Arctic are exerted by winds blowing across the open tundra and creating pressure on the dayboard. The load is reduced by assembling the dayboard with wide spacing between the planks and resisted by adding a substantial arsenal of structural elements such as collar beams and braces.

The field work has revealed that many of the Soviet-era towers remain standing today despite their age varying from fifty to seventy years. This suggests that these structures were more or less efficient, despite being intended as temporary structures that could be easily replaced using both prefabrication and, if required, built on-site with any available local materials.

It is suggested that the weak spot of the timber towers is their foundation. While the later standards regulate this detail of the structure, proposing the usage of concrete pillars and soaking the supports in creosote, it is likely that this procedure was omitted during the construction of the first navigation aids during the 1920s and 1930s due to material and labor shortage. The ruins of towers discovered at mys Krestovskii (Fig. 4) and mys Muksuninskii (Fig. 5) reveal that these structures collapsed due to the disintegration of the foundation elements and main structural members that were attached to them.

Fig. 4. Collapsed tower at mys Krestovskii (dated 1930s).

Fig. 5. Collapsed tower (dayboard lattice) at mys Muksuninskii (dated 1920s or 1930s).

Despite the damp conditions of the environment, the ironmongery remains largely intact, while the wooden members as a result of decades of decade have deteriorated. As seen from the last two figures, the earlier towers were built of roundwood, not beams, as this material was much cheaper (though harder to work with).

3. Conclusion

Navigation on the Yenisei during the twentieth century heavily relied on aids to navigation such as those described in this paper. While many towers continue to stand, most of the Soviet-era structures have collapsed and are rapidly deteriorating. Using primary sources such as the GOST 16150-71, we have been able to classify the tower at mys Krestovskii, though it seems that this structure had been built before the publication of the aforementioned document, which suggests that the later Soviet technical documentation was based on earlier regulations, perhaps composed during the 1950s. The remaining two towers, mentioned in this paper, were built outside the technical regulations of the 1970s and belong to an earlier period. The comparable inexpensiveness and efficiency of wooden structures explains their perseverance into the late twentieth century, when materials such as steel and concrete became available in sufficient quantities.

This paper was not intended as a detailed study on the history of aids to navigation in Arctic Siberia. It merely touched upon a new topic in the field of industrial archaeology and Arctic industrial heritage in order to test its relevance to these fields. The amount of field data collected by the authors of this paper, coupled with scattered data from other expeditions to parts of Arctic Siberia, both professional and amateur, has revealed the importance of this topic to the understanding of maritime and riverine navigation along the waterways of the Northern Sea Route and its adjacent rivers. We hope to be able to produce more comprehensive studies on this matter in the nearest future.

Acknowledgement

This work was conducted within the Eniseiskaia Arktika Project supported by the Foundation of Presidential Grants of the Russian Federation.

References

[1] Tarovik, O V, Kondratenko, A A, Krestyantsev, A B, Topaj, A G 2017 Cargo transportation by waterways of Western Siberia: current status and prospects of development *Vestnik VGAVT* **5** 142–154

[2] Pedersen, N B, Matthiesen, H, Blanchette, R *A et al* 2020 Fungal attack on archaeological wooden artefacts in the Arctic—implications in a changing climate *Sci Rep* **10** 14577 https://doi.org/10.1038/s41598-020-71518-5

[3] Roca, P, Cervera, M, Gariup, G, Pelà, L 2010 Structural analysis of masonry historical constructions. classical and advanced approaches *Archives of Computational Methods in Engineering* **17** 299–325 https://doi.org/10.1007/s11831-010-9046-1

[4] Wilson, S 1944 Early aids to navigation at the mouth of the Mississippi River *US Naval Institute Proceedings* **70/3/493** 279–290

[5] Ryabkova, O V, Kozelchuk, T V. 2017 Ship navigation security of the inland waterways in Yamalo-Nenets National Area during the Great Patriotic War *Vestnik Tomskogo gosudarstvennogo universiteta – Tomsk State University Journal* **423** 175–180 https://doi.org/10.17223/15617793/423/24

[6] GOST 16150-71 Znaki navigatsionnye beregovye. Tipy, osnovnye razmery i tekhnicheskie trebovaniia [Shore Navigation Signs: Types, Dimensions, and Technical Specifications] Giprorechtrans 1972.

[7] Verbist, M, Branco, J M, Nunes, L 2020 Characterization of the Mechanical Performance in Compression Perpendicular to the Grain of Insect-Deteriorated Timber *Buildings* **10(1)** 14 https://doi.org/doi.org/10.3390/buildings10010014

[8] Axisa, R, Muscat, M, Sant, T et al 2017 Structural assessment of a lattice tower for a small, multi-bladed wind turbine *Int J Energy Environ Eng* **8** 343–358 https://doi.org/10.1007/s40095-017-0239-3

AUTHOR INDEX